CREATION
vs.
EVOLUTION
What Was Before We Know?

CREATION VS. EVOLUTION, WHAT WAS BEFORE WHAT WE KNOW?

First edition. April 8, 2024.

Copyright © 2024 Frederick Guttmann.

ISBN: 979-8231780075

Written by Frederick Guttmann.

Frederick Guttmann R.

Thanks to Aday Quintero P. for her collaboration on this and many of my other early ones. I also want to thank Rosemary Garcia and Ira Szczedrin C., as well as my mother Meeky and sister Kersten, who helped me unconditionally with the first revisions of this paper.

Thanks also and blessings to my father Félix and my brothers for their support at all times in parallel with their faith in me, in the face of all the adversities that have occurred in our future.

I dedicate this book to all those who in one way or another yearn for the truth to reach them, and I hope it touches their hearts to remove that veil of ignorance and misrepresentations in which society has plunged us all, leading us after mere hypotheses. and vain philosophies to justify our existence, and that have blinded us so as not to get to know Reality "as it is".

Revision:
Aday Quintero P.
Rosemary Garcia
Ira Szczedrin C.
Maria del Carmen Ramirez
Kersten Guttmann R.
Cover:
Jonathan Guttmann R.
Frederick Guttmann Ramirez, 2012
Web: www.frederickguttmann.com[1]

---

1.   http://www.frederickguttmann.com

CONTENT

Part III.
<u>GENETIC PERFECTION – pg. 145.</u>

<u>THE HEIGHT OF THE FARCE – p. 267.</u>

The best frauds in evolutionary history
The man from Orce
Kanjera Man and Kanam Jaw
The Nebraska Man
The Piltdown Man
· The fraud of a scholar disproves the dating of Neanderthals
· From Australopithecus to Homo Sapiens
· Killings and frauds to justify a non-existent evolution
a lie disguised as truth

Part VII.
<u>THOSE MEN BEFORE ADAM – pg. 279.</u>

· When did modern man appear?
In search of the older woman
The true record
· ...And before the primates there was man
<u>Human evidence out of time - 286</u>
<u>Human artifacts out of time - 296</u>

CONCLUSION – pg. 337.

The last human transition
· To be continue...

*Main sources: 333.

INTRODUCTION

"It is easier to disintegrate an atom than a preconception."
Albert Einstein (German scientist. 1879-1955)

IN THE LAST TWO CENTURIES, a powerful war has begun against all establishments that claim to be theists. Clearly the Catholic occultism of 14 centuries left a strong mark, especially in Europe. The Inquisition, that is, the persecutions, the murders, the arbitrariness, the robberies and the monopoly of Rome still reached the dawn of the Second World War, but the worst damage caused by them was making the world see in Catholicism the representative image of God, and as such, a great hatred was created for everything related to religion. We know that Catholicism was born in the year 325 AD in the days of Constantine, but the religious dictatorship was not born there, because 3 centuries later, another pioneer revived religious wars, putting them at the level of battle lines. Between Islam and Catholicism they led the free world to hate God, because in them they saw reflected the God of which the Bible spoke.

Although the true message of God was lost at the end of the first century at the time of its predecessors, the Messianic Jews or Christian Jews - as they were called in Antioch - had perished. From the first centuries of our era, until the end of the tyranny of Adolf Hitler, the world did not know the history of the Hebrew God, but rather a string of deceptions, misrepresentations, misinformation, ignorance and lies on the part of the institutions that attributed to themselves representation of God on Earth. This clearly led to a lot of anger and resentment, mainly in countries affected by Communism. When glimmerings of new ideas appeared that might contradict Rome, and all the neo-religious movements that emerged between the 17th century and the 20th century, people quickly jumped to embrace them. By the time the End Times begins to approach, another prophecy begins to manifest, thus fulfilling another warning that fits all the pieces determined in the past by the Assembly of God, to bring in the New Kingdom. That oracle says: « *But you, Daniel, shut up the words and seal the book until the time of the end. Many will run from here to there, and knowledge will increase.*" (Daniel 12:4)

The Hebrew prophet Daniel was warned that one of the signs of the arrival of the End of Days would be social bustle, stress, worries, crises, etc., and the extreme development of technological advances, something we know for sure – never better said- which has been happening since the 40s. However, along with said warning, it was also warned that there would be deceit in the world, and one of those areas of falsehood would come from the investigative part: «O Timothy, *keep what has been entrusted to you, avoiding the profane talks about vain things, and the arguments of the falsely called science, which professing some, deviated from the faith.*» (1 Timothy 6:20-21) What is faith? Here enters the continuous struggle between believing what cannot be seen or what can be measured, weighed or observed. Although, dealing with the subject of "God", most think that there is no point of comparison between faith and science, because they clearly start from certain still limited parameters, such as physics and chemistry.

This is due to the fact that we are now postulating new and broader scientific probabilities, study systems such as quantum mechanics, Superstring theory or the Theory of Everything enter the scene, since limited science cannot explain thousands of questions that arise. limited to the material plane. Still, can there be means of confirming Biblical claims? And in this respect we also speak of the origin of man. We will know that when we get into the matter, but despite this, we continue ourselves, limited, starting from two paradigms: evolutionism and creationism. Isn't there more chance? Panspermia can give some ideas of where you can get to study in a more complete field. Panspermia itself is not conclusive, but it serves as a bridge to understand many questions that neither religion nor conventional science have been able to cover.

Thinking that the origin of humanity comes from other parts of the cosmos, or from other dimensions, is what we want to expose in this book, mainly in light of archaeological evidence and professional research by renowned scientists.

"The real problem is not that Darwinism is 'an empty view of life', but a DEFORMED VIEW that turns occasional, even inconsequential facts into fundamental ones."
(Henry Gee, commentator and editor on evolution at Nature)

Part I

SECRET SOCIETIES

THE ILLUMINATI

The Latin term "Illuminati" (enlightened), which appears with the Jewish occultist and former Jesuit, Adam Weishaupt, did not really originate with him in 1776 – the year in which, not coincidentally, the United States of America was also founded -. This name was, and is, used within certain higher levels of Scottish Rite and York Rite Freemasonry. It is also a name to refer to the descendants of the royal houses. And why are they called that? The etymological and historical origin of the word "enlightened" is Akkadian. In the antediluvian era, according to the cuneiform writing of ancient Sumer, tall humans came from space, who called themselves "Anunnaki", the men, and "Antununnaki", the women. Another name they received was Abbennakki, although the Akkadians called them "Ilu" (from where "enlightened" or "illustrious" comes from), that is, in that language, "elevated". The

13

Anunnaki were described by the Hebrews as Nephilim (Fallen), and they were also called Anakim (Giants), in terms of their tall size, a voice that etymologically derives precisely from the Sumerian "Anunnaki".

It is therefore curious that the first Christians - those of the 1st century - who began being mostly Jews, from the sect of the Nazarenes, were recognized among the believers themselves as "Enlightened", for having received the knowledge and truths of the Kingdom. of the Father, of whom Jesus of Nazareth preached. So, "enlightenment", as it could also be understood in Buddhism, can be the state of elevation to an essence, stay or higher status, basically of the mind, through the understanding and comprehension of universal mysteries. Clearly, this knowledge has come more into the hands of evil, greedy, tyrannical and foolish men, than to the humble and meek.

It is wielded from all ancient mythology, that these huge humans were known as the gods of many ancient cultures. They were, citing what was prescribed in the Mesopotamian tablets, astronauts. These came with very specific purposes, among which appeared the objective of having their own lineage on Earth, as the Vedic literature of India declares in detail. They had children with humans, and after generations they placed them in the highest ranks of governments, such as emperors and kings, keeping their blood as pure as possible (the existing literature in this area is quite extensive). From Babylon and Egypt they went to Greece and Rome, then to China and Central America, while the main group settled in Europe, especially in Italy, France, Germany and the United Kingdom (later it reached the United States). They continued to distinguish themselves as "Enlightened" families, as they claimed to have a lineage of celestial royalty, blue blood that comes from the stars, and advice and support that would make them special.

For no other reason, someone important wrote to the Hebrews -its authorship is not really known, although it is presumed that it was Saul of Tarsus-: «*Because it is impossible for those who <u>were once enlightened</u> and tasted the heavenly gift, and were made partakers of it Holy Wind, and likewise tasted the good word of God and the powers of the age to come, and fell away, be renewed again to repentance, crucifying the Son of God for themselves again and exposing him to shame.*" (Letter to the Hebrews 6:4-6, The New Testament) The writer of this letter was considering the Jews who had followed Jesus Christ as "Enlightened". And he adds later: «*But bring to memory the days past, in which, <u>after being enlightened,</u> you sustained a great combat of sufferings; on the one hand, certainly, with insults and tribulations you were made a spectacle; and on the other, you became companions of those who were in a similar situation .*» (Letter to the Hebrews 10:32-33) The Latin version, Vulgate, by Jerome of Estridon, speaks of "Inluminati", referring to those whom Christ told them: "you are the *light of the world*".

The Illuminati movement is known, if anything, more for misinformation and the title stolen by Freemasonry. They believe they are truly enlightened. But why do you think so? When the 30th degree of Freemasonry is exceeded, the members of the order are already aware of who their "god" is, and they were under the orders of a Hidden Master who represents him on Earth, specifically in the brotherhood. One of the joint goals of this hermetic society is the same as that of the satanism of Anton Lavey and Aleister Crowley, which starts from overshadowing any belief in God or Jesus, and misinforming society about the principles of knowledge of the universe. Commonly, today, the 33rd degree Freemasons are called Illuminati, who directly serve a figure called Iblis, to whom the Jesuits and the Vatican leadership report. That Iblis is commonly referred to by high grades as Jah-Bul-On, which some researchers believe to be a trinitarian term for 3 folk gods of the past, including Baal and Osiris.

Recently, leaked information from a Bohemian Grove ritual in California revealed that US presidents and senior British officials participated in ceremonies honoring Moloc, an owl god. The owl and the owl are elemental occult symbols, since the owl or owl (in Hebrew: "lilit") symbolizes the sister of Satan according to ancient Hebrew legends. The very word "lilit" has an Aramaic root, which comes from "laila" or "lilitu", both terms referring to the night or the kingdom of darkness. It is not new to know that the Illuminati are the mass of power that maintains the hegemony of the globe, and are intermingled with the Bohemian Grove, the Ku Klux Klan, the Majestic 12, the Council on Foreign Relations, the Skull & Bones, the Bilderberg Club and in general, the most important clubs or secret societies in the world. For example, the former Prime Minister of the United Kingdom, Tony Blair belongs to the Studholm Lodge, which was founded in 1591, and in the Westminister building itself, where the English Parliament is, there are 2 established Masonic lodges. Another example is that George H. Bush father and son belonged to the Skull & Bones, a secret society from which the leaders of the United States, the FBI and the CFR (Council on Foreign Relations) come out, whose founders were pirates (The Bush family itself is of pirate origin and tied to British royalty).

There are 600,000 Freemasons working across Great Britain, occupying the highest positions of power. Moreover, 1 in 6 police officers in that nation are Freemasons. Why do I give more importance to the United Kingdom? Because that's where the Illuminati came from, who are leading the US today, and there, they came to England, influenced by Germany. Winston Churchill was a member of the Freemasons and the Druids, as it is former officers of the armed forces and the aristocrats who rise to the highest degrees of Freemasonry. Most people do not go beyond grade 3 and do not even know that there are more degrees above the Master's degree. England invaded the US as Spain did with South America, so it is normal that there are currently 4 million people in the US who practice Satanism, according to the FBI. Satanism is organized through high-level Freemasonry, that is, the Illuminati.

The Illuminati are practically the ones who run the world, control it and manipulate it, under the leadership of a world commander that nobody officially knows, but whom they call "Yoda", from where the Freemason George Lucas drew the Jedi character from the War of the Galaxies. In fact, the Jedi in Freemasonry are demons. And where did they get all this? from Egypt and the Jewish Kabbalah, that is, taking the deep information of Solomon and adopting it for evil, for occult purposes. This was brought to the UK, where the power of the Zionists is, the dark Jews who control the world monopoly. For this reason they are called 'British', related to the Hebrew "brit", which is "covenant", and "ish", which is "male". They are "the man of the pact" of Satan, since they are considered the cradle of the Antichrist, according to documents and testimonies of Illuminati leaders.

This is why they control the world's oil (BP = British Petrolium) and the economy (WB + FR = World Bank, or World Bank + Federal Reserve and IMF = International Monetary Fund). The owners of the other oil companies like Shell are the Bushes. This subject is extensive, but we have left the reader a rough smattering of some of these intriguing truths, but if you want to go into more detail about it you can read in more detail the works of William Bramley, Milton William Cooper, RA Boulay, Alex Collier, David Icke, Alex Jones, Eric Jon Phelps, Alexander Hislop, Sixto Paz Wells or Michael Tsarion. We also deal with these themes in our works on the End Times.

FRATERNAL ORDER

Freemasonry plays a transcendental role in this whole story, since they took the role of establishing atheism at all costs, although the newly initiated, as in Catholicism, know little or nothing about this, nor about the levels of satanism that exist. beyond the 33rd degree of the Scottish Rite. Mason means "mason" or "builder" in English. From ancient Egypt come the ideas that today are ventilated in Freemasonry, which intensified with the mystical literature that King Solomon left written. Since the beginnings of classical Greece, this information had been passed on from Egyptian priests to Greek philosophers, and it was very regular within royal families that distinguished themselves by saying they had "blue blood." With the death of Alexander the Great, the Roman Empire began its ascension, continuing the traditions and the secret monarchies, which later rose to Germany, France and England, settling in the most secret until today. Over the centuries secret societies, inspired by the Mystery School of ancient Egyptian priests, moved within Europe, and later moved to the US.

These groups maintained themselves as fraternal groups, private clubs, hermetic orders and mystical societies, which strove to study and revive the great enigmas of occultism and universal knowledge. The fraternities were later called Lodges - from which the word "logic" is used - and were identified with the letter "G" (gnosis = knowledge, from the Greek word that became English "know ledge"). Charles Darwin, recognized revolutionary of the theory of the origin of man by processes of mutation of species, some towards the immediately superior ones, was the pawn that Freemasonry urgently needed to achieve social sovereignty over religion. His most famous literary works were "The Evolution of Species" and "Natural Selection", and which officially gave birth to the commonly called "Theory of Evolution". Charles Darwin, with the approval of Freemasonry, radically changed the perception of man's past through proposals that, despite always having been treated as "hypotheses", have today become an approach accepted and taught as "a resounding fact"."

This set of speculations supports, as an absolute statement of the origin of our human race, the advent of a large number of chance events that have led the strongest to survive and improve. The international scientific community is based on this ideology to say that man comes from monkeys, and that these, the primates, in turn, come from a type of prehistoric shrew that would have survived underground after one of several mass extinctions, and before this mammal, we would have descended from other animal species that evolved from dinosaurs. Traveling further back in this evolutionary order, humans would have arisen from invertebrate organisms, and before, from cells that were the first to begin to adapt to their environment, and give this order of mutation, one to the other to become organisms. more complex.

As can be read in any book that emphasizes the biography of Charles Darwin, he was born in Sherewsbury, England in the year 1809, at the age of 8 he was orphaned by his mother, his father being Robert Darwin, who took care of the offspring made up of her Erasmus siblings, Mariannne, Carolinne, Catherine, Susan and Charles himself. His father was a village doctor, passionate about the study of science, a hobby that Charles undoubtedly inherited, and that Robert would have acquired from his father Erasmus. In 1825 he entered the University of Edinburgh to study medicine, but upon learning of Lamarck's evolutionary hypotheses, he turned his studies towards zoology, not without some displeasure on the part of his father who recommended that he dedicate himself to an ecclesiastical career. In this way, he began his studies in Theology in Cambridge (1828), obtaining the title of Bachelor of Arts, the only title of his life. At this stage he attended the botany classes given by Stevens Henslow, from whom he received a strong influence, since it was thanks to this teacher that Darwin developed many of his ideas. In addition, he receives his knowledge in entomology, botany, chemistry, mineralogy and geology.

The same professor encourages him to enlist without pay on the Beagle frigate, organized by the British Admiralty to travel through the Southern Hemisphere, with a half-scientific and half-commercial mission.

TRAVELING ON THE SEA

Charles Darwin was 22 years old when he embarked on the Beagle, and in the Galapagos Islands he had an encounter that was surprising to him, upon discovering, in a territory far removed from the South American coast, some giant lizards that science at the time believed to be extinct. The important thing for him was to discover that totally different species existed on other nearby islands, despite the fact that the climate and orography did not show any variation. The explanation that began to take shape in Darwin's mind was that a particular adaptation of the species to the specific conditions of their habitat had to take place, in a process that would last millions of years and in which the same species would have suffered changes and mutations that would allow it to develop a more efficient adaptation, bearing in mind the limited range of scientific branches that existed at that time. His conclusions seemed new, however, he already came from the "school" of his grandfather Erasmus Darwin, and his approaches were just another one from that English era of new philosophies, where many others already believed and defended hypotheses of evolutionary processes and ideas. of chance. Believing in these things and establishing them publicly was a rather hard social blow, since it was a time when the religious explanation, based on Catholicism, regarding the origin of the world, was completely unquestionable and was the official opinion.

Darwin took daily notes and recorded every piece of evidence he found of strange fossils and then unknown species. All the information that he was able to collect during his research led him to his evolutionary hypothesis, which he expressed in his book "The origin of the species through natural selection". This work came to light in 1859 and it explained, apparently in a scientific way, the origin of the human being. Not only did he remove the role of "Creator" from "God" but he firmly insisted that Nature was "self-generating" - a point of view that was obvious to society at the time, but centuries later was seen to be wrong. a huge inconsistency. From that very moment, Darwin began to dedicate himself to writing articles and giving lectures, placing himself in the focus of an uncomfortable polemic regarding the representatives of the Roman Catholic religion and traditional science. This hypothesis was completed and finished in 1871 when his second book was published, "The origin of man and selection in relation to sex."

The firmest and most solid foundations of an entire explanation of the world and of man, which was based on a genesis narrative, began to shake. Poorly translated narratives such as those that we can see today in the famous and ancient book of the Bible: Genesis, began to be questioned, bringing with it a great controversy. Upon returning to England on January 29, 1839, Charles Darwin married Emma Wedgwood, his cousin. They both decided to change their residence and moved to Down in 1842, near London, already being the parents of ten children, all the fruit of their marriage. Darwin drastically reduced meetings, conferences and, in general, appearing in public, using excuses regarding his health, due to his dedication to the study of intellectual creation and the writing of his thought, a study which would later bear fruit to his son Francis who published his work with the title of "Memoirs of the development of my thought and my character". The contribution of Charles Darwin's ideas that species are not immutable coincides with what was exposed by a contemporary of his, Alfred Russell Wallace. Both presented their theory in 1858 at the Linnean Society in London, causing the logical convulsion.

THEORY AND HYPOTHESIS *of forced origin*

Before taking Charles Darwin's hypothesis as a "theory", we should keep in mind that a "theory" is " *the set of reasoning devised to provisionally explain a certain order of facts.*" Darwinism years ago fell under its own weight. However, these concepts continue to be officially taught in most countries and continue to be the central focus of the vast majority of modern scientists, under a common tendency towards anthropocentric defense. However, science, as the study of things, must not be tendentious or subjective: « *Evolution is not even a theory, but a hypothesis, which, in the "scientific method" is one step below becoming a theory. theory.*" (Carl Baugh, Ph. D. "The Dinosaur Dilemma", 1994. Texas, USA). Now, if we remember what a "hypothesis" is, we see that it is an " *assumption that a thing is possible in order to draw a consequence from it.*" Something more akin to Darwin's postulates. These approaches finally include only assumptions to have a provisional explanation of the true appearance of everything, so we have to refer to such as "hypothesis", mainly because the archaeological evidence does not support this popular paradigm.

Now, dealing with the important point of the core, Charles Darwin did not directly belong to Freemasonry -or so it is believed-, but it must be agreed that he was always closely linked to the brothers of the Order and that, without the active presence and their influence, the young man's life would have been completely different. It was his grandfather, Erasmus Darwin, who got him into all the deliberate mess of Freemasonry and its anti-religious aims. This famous physicist and botanist, a Freemason initiated in 1754 in the Lodge of Saint David No. 36 in Edinburgh, greatly influenced his grandson's decisions. To such an extent that he used to call him "Father Erasmus". Nor would it have been possible for the enthusiastic researcher to be part of the Beagle passage had it not been for the intervention of another Freemason. It so happens that, as Charles had barely passed twenty years of age, his father Robert Waring Darwin was strictly opposed to the projected trip. For this reason, he stated that he would only change his mind if "someone with common sense" persuaded him that the long journey was the right thing to do. That someone existed.

It was his uncle - and future father-in-law - the freemason Josiah Wedgwood who, with his intervention, managed to make his young nephew achieve such a long-awaited goal. At the same time, it must be remembered that Dr. Thomas Henry Huxley, a Freemason member of The Royal Society, was the one who was most concerned with getting Darwin to write down his thoughts on the supposed evolution of the species. Huxley later became what we would today call Darwin's "official spokesman," or "Darwin's Pitbull." Between Darwin, Huxley, John Hooker, Herbert Spencer, they founded the X Club in 1864, together with a series of powerful scientists - its head was Huxley, member of ten royal commissions, president of all geological societies, the Linean society and so on. scientific power movements in the UK. The goal of this X Club was to promote Darwinism. For these, they were the mixture of the best of the races, and the blacks and Hispanics were the lowest and what should be eliminated. Even with everything, « *science does not explain everything, they are approximations to the truth*» (Máximo Sandín) And this professor adds: «*Darwin was an amateur naturalist, not a professional scientist. His graduate was theology. Nor was he distinguished by brilliance in his exhibitions* » (Máximo Sandín)

In his biography, Darwin himself acknowledges this. Darwin based his concept of Natural Selection on the observation of domestic animals, but nature cannot be understood in the same way as the development of animals in captivity or in common areas with people. « *For Natural Selection to exist, there must be characteristics that are recorded in our genes, because if not, there is nowhere to select from.*» says Sandín, concluding that then " *we can't talk about evolution... actually we should call it Transformation .*", because we talk about evolution as if we were mentioning a Marvel Comics mutant fiction novel, where an alteration makes humans superheroes.

THE FAMILY MASONIC Mission

"*... a century and a half before Darwin, science was not separate from religion, but rather an aspect of religion and essentially in the service of it...Darwinian science therefore became represent a great threat not only to the theological claims of religion but also to the functional usefulness of religion [which] confers purpose and meaning.*" (Michael Baigent, Richard Leigh, Henry Lincoln) The publication of Charles Darwin's book, "The Origin of Species", represented a very critical high point in Freemasonry's war against religion. Actually, in the book there is no information about the "origin of the species", but this did not prevent it from becoming popular in a very short time, since the real purpose was not "scientific" but to be ideologically victorious. Much less would it cease to have a place in the strongest years of Materialism and Marxism in the United Kingdom to be the ideal bridge for atheism. The ideas that Charles Darwin proposed were not born in his head, but rather came from his teachers, his family and the books he constantly read, such as "Principles of Geology" by Charles Lyell, who used the studies of James Hutton, also Geology buff.

According to both, the world was not as young as the Catholic Church and Islam made it out to be, which many people believe is what the Bible says. A misinterpretation of the Old Testament

Genesis says that the world and everything in it was created 6,000 years ago, but clearly Genesis speaks of 7 days, which is to be understood as representing a larger but unknown number. However, Hutton and Lyell claimed that the Earth was billions of years older than monotheism claimed, something completely unacceptable by the Christian and Muslim world. Obviously, and due to the lack of knowledge and culture in the scientific field at that time, there was no one to present a counter-investigation to destroy said thesis.

The Father Erasmus

Charles Darwin's grandfather, Erasmus, was the one who motivated him to become irreligious. Charles Darwin had been in the habit of listening to his grandfather since childhood and was also greatly influenced by his opinions. Erasmus Darwin was virtually the first person to put forward the notion of "evolution" in England. The most important characteristic of Erasmus Darwin was being one of the few precursors of "Naturalism" in England, a trend of thought that assumed that the essence of the existence of the universe was in nature, while denying a metaphysical creator, and considered as the Creator to nature itself – even though nothing can create itself. In other words, it was a variation of the materialistic thinking that dominated the 19th century. The naturalistic studies previously concluded by Erasmus Darwin completely paved the way for Darwinism. On the one hand, he had carried out investigations of a botanical nature in a garden that he owned two acres from where he drew arguments that would form the main elements for Darwinism, investigations which he compiled and shaped in his books "The Temple of Nature" and " Zoonomy".

On the other hand, in 1784, he had created a society that would pave the way for spreading these ideas: the Philosophical Society. It is not surprising that later, some ten years later, after the establishment of said society, it became one of the largest and most effusive supporters and defenders of the theory expounded by Charles

Darwin. The conclusion is that, quite possibly without even knowing it, Charles had become the puppet of a new system carried out by Freemasonry with the aim of abolishing religion. Erasmus Darwin had another very characteristic feature: he was the representative of Freemasonry. This was the sovereign and main founding force of the New Secular Order or New World Order (New Order for this Century, eliminating religions and governments), which reached a very high point in the 19th century and which today leads the world to the globalization. Dean Darwin was one of the teachers of the well-known Canongate Kilwinning Lodge in Edinburgh (Scotland).

He was also connected with the Jacobin Freemasons of France and with the Illuminati society, which had made anti-religious work its main task since its inception. Erasmus had raised his son Robert (Charles's father) in the same thoughts and enrolled him in the Masonic lodges. Because of this, Charles Darwin was to receive a Masonic inheritance from both his father and grandfather. Undoubtedly, this carries an important meaning because Freemasonry was one of the central powers that led the long fight to bring down the socio-economic order that was based on religion and replace it with a secular order as seen today already prevailing in all countries. economic, political, military, informational, corporate and social sectors.

As we have defined above, there was only one aspect missing from the New Secular Order gala, that is, the presentation of a non-religious explanation for the existence of all living things. In reality, what Charles found was nothing more than a worthless argument, a chimerical statement impossible to be verified by means of solid and true evidence. Conversely, it was a claim prone to permanent refutation. But this situation would not lead it to lose value in the eyes of the New Secular Order, which would take advantage of it as its best weapon against God and would establish it first in society as a consolidated idea, and whoever refuted it would be seen as ignorant. (Source: The Deception of Evolutionism, by Harun Yahya)

THE LEGACY

Erasmus Darwin, being a representative of Freemasonry - today commonly called "The Church of Satan" by many -, would be named in the Illuminati degrees. The Illuminati, and Freemasons in general, are very similar to political parties: they sell a propaganda image and then behind the theatrical curtain they have a few drinks. Freemasonry, or freemasonry, is identified as an institution of an initiatory, philanthropic, philosophical and progressive character, founded on the feeling of fraternity, equality and freedom. As a brand or product, they present themselves by saying that they have as their objective the search for truth and promote the intellectual and moral development of the human being, as well as social progress. Freemasons, both men and women, are organized in base structures called lodges, which in turn can be grouped into a higher-level organization usually called "Grand Lodge", "Grand Orient" or "Grand Priory". In addition to this, Freemasonry is the main power in conducting the intellectual changes necessary to distort the spiritual order, using various mechanisms to do so. Freemasonry had won a considerable victory over the Catholic Church thanks to the alliance established by anti-Christian forces. The 19th century was the gala of the New Secular Order instituted through that victory and which still fights today to destroy the Vatican and enslave the human race. A related example would be the movie: "Angels and Demons" by Dan Brown.

NATURALLY

Common phrases such as "naturally", "nature is wise" or "nature knows what it does", have changed the perception of God as coordinator of the things created by him, to give the role to the goddess Gaia (Gea), without society itself realizing it. Now, what was the Theory of Naturalism, accepted only what was perceived in nature and through the senses. Nature was considered to be the creator and ruler of itself. Macho formulations such as " *nature created women to be in the way* ", are common manifestations of the mentality injected into society by the naturalist movement, which generalized expressions such as "Mother Nature" or "Nature". Naturalism was promoted by a well-known organization: Freemasonry.

It appears again on our map, and it is especially proclaimed in the well-known encyclical of Pope Leo XIII (1810-1903 AD) Humanum Genus: «In our *time, with the help and support of a society called Freemasonry, which possesses a broad and effective organization, the efforts of those who worship obscurantism have joined. They no longer feel the need to hide their ill will and fight against God Blessed* ". That Pope disclosed the relationship between Naturalism and the Masonic organization: "*All the objectives and efforts of the Freemasons lead to one intention: to abolish all religious disciplines and social principles of Christianity and establish a new system of norms based on the principles of naturalism and on his own ideas.*" Charles Darwin was the one who indisputably contributed the most to Naturalism, including the fact that he sealed the door on the possibility of demolishing this theory. Those who firmly believed in Naturalism and felt true admiration and intrigue for the perfection of nature, saw how their exciting theory became obsolete by not finding an answer to the question of "who or what gave it life" or "who or What made her so perfect?

However, they flatly refused to accept the existence of a God, responsible for everything, since their positivist approach forced them to simply believe in the concepts that, as a result of experiments and observations, take shape. This implies, neither more nor less, that nature has created itself, making itself the creator of everything known up to now, including us. But let's keep in mind that this theory is absurd since nothing can create itself without more. Obviously, this was what Darwinism wanted to change. His claims constituted a "foundation" for the claim that nature created itself. The criterion of natural selection states that the weak individuals of a species are eliminated in the struggle for life and that the strong ones that remain are responsible for the improvement of that particular species. Perhaps this explanation is not wrong, but Darwin himself did not make use of this process properly, although it is not feasible if it is supported by the theory of "favorable mutation".

The only thing that natural selection could achieve was to make certain individuals stronger, for example, to survive. In other words, natural selection could only be responsible for the improvement of generations. Even so, "the origin of species", which was the name of Darwin's book, could never be explained by means of natural selection. This is so because natural selection does not transform a donkey into a crow or a dolphin into a rhinoceros. These species were created separately and natural selection could well be responsible only for the elimination of the "weak" individuals and the survival of the more gifted.

LAMARCK

In short, Darwin's ridiculousness even began with the title he gave to the book, which would talk about "the origin of the species" and where he did not explain, in any case, said "feat." On the other hand, we must admit that in the time of Charles Darwin there was a rather large deficit in terms of knowledge of biology and therefore it is not surprising that this character's hypotheses did not seem so fanciful, even though they were considered valid. The more time passed and the more data and information was collected, the more evident the contradiction in Darwinian philosophy became and, of course, the more it was kept secret. It should be noted that he was also completely unaware of certain genetic distinctions since, after a revision of his original words, he came to propose another theory in which he said that a subspecies of bees fed more and more on animals that lived in the water, thus developing their mouths being longer and longer giving rise to whales, something completely logical in the fabulous theory of evolution.

On the other hand, biology and geology grew in terms of knowledge and discoveries and followed the same path, separating more and more from Darwinian theories. Meanwhile, Darwin insistently affirmed that the Earth was approximately 300 million years old, and that figure is not even 10% of what is calculated today in terms of the real age of the Earth. The ratification of naturalism, even by means of deceptive methods, was very important due to its socio-political consequences. The promoters of such a New Secular Order accepted the social and individualistic models generated by it, and explained nature using them. It was based on these models because it thus "demonstrated" that the New Secular Order was also the order by which nature was governed, fully reflecting its characteristics. This was one of the triumphs achieved by Darwinism in the name of the New Secular Order.

The French naturalist, Jean-Baptiste Lamarck, thought that living beings passed on the knowledge they gained to their future generations, but he also thought that if arms or family members were cut off for many generations, babies would begin to be born without them. Darwin was greatly influenced by these examples and even added new arguments. In "The Origin of Species", he states that some bears, which have been in the water for a long time, evolved into whales. But Lamarck and Darwin were wrong because their ideas contradicted the fundamental laws of biology. In their time, they did not have genetics, microbiology or biochemistry as fields of science (Source; documentary: "Evolucionismo, El Engaño Materialista"). "*We know that any science must have its philosophy and that only in this way does it make real progress [...] if the philosophy of science is neglected, its progress will not be real and the entire work will remain imperfect.* " (Jean-Baptiste Lamarck. 1744-1829)

The ideas of Jean-Baptiste Lamarck were not taken into account in his time, although his book "Zoological Philosophy", where he expressed his theory, circulated in France and also in England, a work to which Darwin himself had access (who later published it). attacked in his treatises). During the 20th century, Lamarckism has been defended by different evolutionists, and the one known as the "Baldwin effect" (announced by James Marck Baldwin and C. Loyd Morgan at the end of the 19th century), a sweetened version of Lamarckism according to which sustained habits of species, by natural selection, would be fixed in heredity, it is maintained as plausible to solve some difficulties of neo-Darwinism. Lamarck actually formulated the first theory of evolution.

He proposed that the great variety of organisms, which were static forms created by God, had evolved from simple forms; postulating that the protagonists of this evolution had been the organisms themselves due to their ability to adapt to the environment. Renowned biologists such as Lynn Margulis and Máximo Sandín consider that « *a main suggestion for the new century in biology is that the maligned slogan of Lamarckism, "the inheritance of acquired characters", should not yet be abandoned: it should only be carefully refined.*" Inquiring into history, it can be evidently concluded that Charles Darwin was a genius "manufactured" by the powerful interests of his time, being the son of an extremely wealthy doctor.

NATURAL SELECTION

As a process of nature, natural selection was familiar to biologists before Charles Darwin, who defined it as " *a mechanism that keeps species unchanged without being corrupted.*" Charles Darwin was the first person to affirm that this process had evolutionary capacity and later mounted his hypothesis on that basis. The name he gave his book indicates that natural selection was the basis of the theory: "The Origin of Species Through Natural Selection." However, since Darwin's time there has not been a single shred of evidence to show that natural selection causes living things to evolve. Colin Patterson, a paleontologist and dean of the Natural History Museum in England, a prominent evolutionist, emphasizes that natural selection has never been observed to have the power to make things evolve: "No one has ever produced a species through the *mechanisms of natural selection. No one has ever come close to it, while most current neo-Darwinian arguments deal with this question.*" (Colin Patterson, "Cladistics", interview with Brian Leek, Peter Franz, March 4, 1982, BBC)

Natural selection contributes nothing to The Theory of Evolution because it can never increase or improve the genetic information of a species. Nor can it transform one species into another: a snail into a fish, a fish into a toad, a toad into an alligator, or an alligator into a bird. The greatest advocate of punctuated equilibrium, Gould, refers to this insurmountable discrepancy of natural selection: " *The essence of Darwinism lies in a single sentence: natural selection is the creative force of evolutionary change. No one denies that natural selection played its part in eliminating the unfit. (But) Darwinian theories require that it also originate the expedient.*" (Charles Darwin, The Origin of Species: A Facsimile of the First Edition, Harvard University Press, 1964, p. 189)

TRICKS AND FRAUD METHODS

According to Professor Harun Yahya, writer of the book "The Deception of Evolutionism", another of the deceptive methods that evolutionists also use in the matter of natural selection, is to present this mechanism as if a conscious designer were at work. However, natural selection does not have any kind of consciousness. It does not have the will to decide what is good and what is bad for living things. Consequently, natural selection cannot explain biological systems and organs that have the character of "irreducible complexity." These systems and organs are made by the cooperation of a large number of parts and are useless if one of those parts is missing or defective (for example, the human eye does not work unless its constitution encompasses all the details that make it work). make fit for vision). Therefore, the will that brings together all the parties to the case should be able to imagine the future in advance and aim directly at the benefit that has to be acquired at the last stage.

Since the mechanism of natural selection does not possess any consciousness or will, it cannot do any of that. This fact, which also demolishes the foundations of the theory of evolution, also tormented Darwin: « *If it could be shown that some complex organ existed, which, perhaps, would not have been formed by means of numerous, successive and slow modifications, my theory would absolutely collapse.*" The so-called natural selection only separates the deformed, weak or inept individuals of a species. It cannot produce new species, new genetic information, or new organs. That is, it cannot make something evolve. Charles Darwin accepted this reality saying: "*Natural selection can do nothing until favorable variations occur by chance.*" (The Origin of the Species. Page 177). This is why neo-Darwinists have had to present mutations, contiguous to natural selection, as "the cause of beneficial changes."

However, mutations can only cause "harmful changes" but never favorable ones. « *The highest representatives of science see themselves as locked in a desperate battle against religious fundamentalists, a label they tend to apply generally to anyone who believes in a Creator who plays an active role in the affairs of the world. Darwinism plays an indispensable ideological role in the war against fundamentalism. For that reason, scientific organizations are dedicated to protecting Darwinism rather than testing it, which is why the standards of scientific inquiry were shaped to succeed in doing so."* (Philip E. Johnson, University of California) The evolutionary scenario that life arose from nonliving matter by chance is rejected by present-day scientists. Moreover, there is no mechanism of nature to support the mentioned process called "evolution". There is no natural mechanism by which a simple cell can transform into a more complex creature (Source: documentary "Evolucionismo, El Engaño Materialista").

IT'S WAR!

Where exactly did the accepted paradigm come from? To speak properly about the evolutionary hypothesis we must refer once more to the time of Erasmus Darwin, Charles's grandfather, and again to the war between Freemasonry and the Catholic Church, likewise, to the links between the Darwins and the Freemasons. in favor of abolishing any belief in a supreme God. Conventionally, people follow canons continuously, whose origins they don't even know and often defend them tooth and nail. This is a topical idiosyncrasy of society, where it is more comfortable to "receive chewed food." People don't investigate the sources of things. They simply follow others who profess things that others say, and so on, the chain continues. It is like the group of scientists who carry out an experiment in a room with several monkeys. These monkeys see a banana hanging from the ceiling – they don't know it, but it's not a real banana, and it's connected to electricity – and just below is a metal ladder.

One by one they climb the ladder and touch the banana. Each one has to go up to go through their own experience, but it turns out that they introduce a new group of monkeys into the room that know nothing about the banana connected to the current. The new group tries to go upstairs to get the banana and the old ones start jumping and shouting, so that the old ones, without knowing what the problem is, simply don't go up the stairs. Human beings act in the same way. In 1776 a striking society was founded in southern Germany, in Bavaria. The founder of this society called "Illuminati", that is, "Enlightened", was the law professor, renegade Jacobite, Freemason and occultist Adam Weishaupt. Said society had two attributes that made it very interesting to people denied to God: it was a satanic sect and it established a very pretentious political program, written by Weishaupt, where the main purpose was defined as follows:

1. The abolition of all monarchies and governmental systems.
2. The abolition of all "theistic" religions.

This society was extremely opposed to religion. According to what the English historian Michael Howard expresses, Weishaupt felt a "pathological hatred" towards divine religions. In truth, said society was one more Masonic lodge, given to potentially instilling Darwinism. Very few of the lodge-goers could see the "grand master" Weishaupt face to face. In 1780, with the participation of Baron Von Knigge, one of the greatest masters of the German Masonic lodges, the power of the society was further expanded. Weishaupt and Knigge began preparations to make a revolution that would be defined as "socialist." However, when the government discovered this undertaking, the enlightened masters, Weishaupt and Knigge, decided to participate only in the ordinary activities of their lodges and dissolved the aforementioned society. This step was taken in 1782, at which time other branches of Freemasonry were also establishing themselves in England under the atheist banner.

MARX

In the early years of the 19th century, a new society was established in Germany that sought to preserve the "Illuminati" tradition. That society became known as the "Honest Men's Association." After some time it changed its name to the "Association of the Communists". Karl Marx (father of Marxism) and Frederick Engels wrote the Communist Manifesto in accordance with the instructions received from the last named association. As is known, said Manifesto defined religion as "the opium of the people", assuring that one of the conditions of an ideal society should be a "classless society", at the same time that it considered that the only path for the salvation of humanity was the elimination of all religious beliefs. In reality, the "Illuminati" Freemasonry society founded by Weishaupt, and its extension, the "Honest Men's Association", were two more similar Masonic organizations that were established in Europe in the 18th century.

The common feature of all of them, in parallel with the Enlightenment philosophy dominant at the time, was their vigorous opposition to monotheistic religions. Since the Enlightenment philosophy imposed the idea that the only guide for human beings was their own reasoning, it was proclaimed that Divine guidance or divine inspiration was not needed at all. The most important political consequence of the Enlightenment was the French Revolution. The most evident feature of this was the hatred of the Church and, even more, against religion itself. In the most chaotic days of the Revolution, a broad movement to "get rid of religion" developed as a result of the intense propaganda of the Jacobins, pioneers of the coup. In addition, the people were introduced to a new "religious spirit." Consequently, the "cult of the Revolution" began to spread through propaganda, which was witnessed for the first time at the Festival of the Federation that took place on July 14,

1790. Robespierre, the known leader of La Revolución, presented some norms for this "revolutionary cult".

He defined the maxims of this worship in a report that he called "Cult of the Supreme Being." One notable consequence of this was the transformation of the Church of Notre Dame into the "temple of reason." Therefore, the images located on the walls of the church were lowered and a statue of a woman was placed in the middle of the building, which was defined as The Goddess of Reason. The first French Revolution, in 1789, marked the beginning of a long series of rebellions in France. The new Duke of Orleans, Louis Philippe, became the figurehead of a revolt in July 1830, which placed him on the throne of France as ruler of a constitutional monarchy. Assisting him was the Marquis de La Fayette. Another important supporter of Luis Felipe was a man named Luis Augusto Blanqui, who was decorated by the new ruler for having helped make the revolution of 1830 a success. Blanqui continued to be an active revolutionary after 1830 and provided significant leadership to a long chain of of rebellions.

According to Julius Braunthal, writing in his book, "History of the Internationale": « *Blanqui was the inspirer of all the rebellions in Paris from 1839 to the Commune in 1871.* » Blanqui belonged to a network of secret societies in France, which organized and planned revolutions. Almost all of these secret societies were offshoots of the activity of the famous Masonic Brotherhood, and were constituted in the image of their organizations. Each society had a different function and ideological basis for recruiting people into the revolutionary cause. Although revolutionary societies sometimes differ in ideology and tactics, they had one goal in common: to bring about revolution. Many revolutionary leaders participated in several of those organizations simultaneously. One of the most effective secret groups of French revolutionaries was the Society of Seasons, of which Blanqui shared the leadership.

This society was explicitly planned for the purposes of hatching and carrying out political conspiracies. One of the organizations allied to The Society was The League of the Just. Years after this uprising, The League was joined by a man who later became its most famous revolutionary spokesman: Karl Marx. Karl Marx was a German who lived from 1813 to 1883. He is considered by many to be the founder of modern Communism. His writings, especially the Communist Manifesto, are an important cornerstone of communist ideology. However, as some historians have claimed, Karl Marx did not originate all of his ideas. He acted for some time as a spokesman for a radical political organization to which he belonged. It was during his membership in the League of the Just that Marx wrote the Communist Manifesto with his friend Frederick Engels. Although the Manifesto contains many of Marx's own ideas, its real achievement was putting into coherent form communist ideology, which was already inspiring France's secret societies to rebellion.

This was an attempt to recover the Illuminati ideology of the occultist Adam Weisaupt with the Association of Honest Men in the 19th century (it was at that time Karl Marx (founder of Marxism) and Frederick Engels wrote the Communist Manifesto. For them religion was the opium of the people and believed that the salvation of humanity would be the elimination of all religions). Both Marx and Erasmus were fundamental pieces in this merciless war. Due to his intellect, Marx gained considerable power within the League of the Just and his influence caused few changes within this organization. Marx did not like the conspiratorial-romantic character of the network of secret societies to which he belonged, and managed to eliminate some of those traits within The League. In 1847, the name of La Liga was changed to the "Communist League". Associated with the Communist League were various workers' organizations such as: The German Society of Educational Workers GWES. Marx founded a branch of the GWES in Brussels, Belgium.

From these workers' societies we see the rise also of Adolf Hitler almost a century later. At this point, we can see the extraordinary irony in those events. The same network of Masonic Brotherhood organizations that gave birth and power to the United States and other capitalist countries through revolution, were now actively creating the ideology that opposed these countries: communist ideology. It is crucial that this point be understood: both sides of the modern struggle—communists versus capitalists—was created by the same people in the same network of secret Masonic "brotherhood" organizations as "Machiavelli's method". This vital fact is almost always overlooked in the history books. Within a short period of one hundred years, the network of "the Brotherhood" gave the world two opposing philosophies that provide the entire foundation for the so-called Cold War: a conflict that lasted for nearly half a century. (Source: William Bramley in "God's of Eden")

The relationship between secret societies, occultism, satanism, atheism and Darwinism must be clear: if we do not know where we come from, we cannot know where we are going. The same happens with our beliefs: we must know where they come from and for what reason they have been promoted. Some of them have disguised themselves very discreetly with humanism and pro-religion masks, when deep down their interest is rooted in the objectives of the New Secular Order. Marxism is highly apocalyptic. It tells of a Final Battle, involving the forces of good and evil, followed by a utopia on Earth. The primary difference is that Marx molded this idea into a non-religious framework and tried to make it sound like social science rather than religion. In Marx's scheme, the forces of good are represented by the oppressed working classes, and evil by the propertied class. Violent conflict between the two classes is depicted as natural, inevitable, and ultimately healthy, as such conflict will eventually result in the rise of a utopia on Earth.

Marx's idea of the inevitable class struggle reflects Calvinism's belief that conflict on Earth is healthy, because it means that the forces of good are actively fighting the servant of evil. Also, what direct relationship did Karl Marx have with Charles Darwin? Darwinists claim that Karl Marx sent his literary work to Darwin, although he did not even read it, but they do not add Marx's own views on Darwin, who later wrote to Frederick Engels: "As for Darwin, to whom I have read *another At the same time, I am amused when he tries to apply Malthus's theory equally to flora and fauna, as if Mr. Malthus's cunning did not reside precisely in the fact that it does not apply to plants and animals but only to men – with geometric progression - as opposed to what happens with plants and animals. It is curious to see how Darwin discovers his English society in beasts and vegetables, with the division of labor, competition, the opening of new markets, "inventions" and the "struggle for life" of Malthus. "* (Karl Marx)

⸺ ⟲ ⸺

MASONRY, OR FREEMASONRY

What is Freemasonry itself? Freemasonry defends that it is a brotherhood or circle of wisdom where they keep and guard great knowledge. But where does it come from? Freemasonry was inspired by the legendary Egyptian Mystery School where the gods gave great knowledge to the priests and pharaohs and they kept them more and more jealously, to the extent that they no longer allowed anyone access to knowledge. In this way, hermetic groups or secret societies follow this pattern as the first rule. According to Dr. H. Spencer Lewis, founder of The Rosicrucian Order Headquarters in San Jose, California, the first temple built for the use of The Mystery School was erected by Pharaoh Cheops. Within that temple, spiritual knowledge suffered the deterioration that caused the pharaohs to mummify their bodies and bury wooden ships. According to information from ancient Egypt, the distorted teaching of The Mystery School was created by the "great teacher" Ra.

We also see initiation levels, similar to those of ancient Egypt. The word freemason comes from English: "free-mason", which means "free-mason", and this sect has a staggered order in 3 phases: apprentice, official and master. This arises from the Egyptian Book of the Dead dated from the year 1591 a. C. We have in the same order the 3 important symbols of Freemasonry: a ruler, a square and a mallet. But where do these icons come from? That is answered in the classic novel by the Master Masons that talks about Hiram Abiff, king of Tire in Phoenicia – a descendant of the Israelite tribe of Naphtali -, as the Bible says, he helped King Solomon to build the first Temple in Jerusalem (1 Kings 7:13).

According to the MM.MM (Master Masons) one night Hiram was transported to the underworld and the last descendant of Satan, named Tubalcain (Genesis 4:22) told him: «At the *beginning of time, there were two gods who shared the Universe, Adonai, and Iblis* (also called: Samael, Lucifer, Prometheus, Baphomet), *the master of fire. The first created man out of clay and animated him, and then he created Eve. It was then that Iblis sent his sister Lilith and made her Adam's lover, while Iblis seduced Eve and impregnated her* (according to Talmudic traditions, Cain was born from the love of Eve and Iblis, and Abel from the union of Eve and Adam). *Later, Adam will feel nothing but contempt and hatred for Cain, who is not his true son. One day, Cain, tired of seeing ingratitude and injustice, will rebel and kill his brother Abel.* »

After this vision Hiram would return to the surface stunned. Almost finished the works of the Temple of Jerusalem, three companions who found it difficult to be admitted to the master's degree, decided to obtain it by force. Each stationed at a door of the Temple, they invited Hiram to reveal their secrets. As he did not want to reveal them, each one dealt him a blow (one with a ruler on his throat, another with an iron square on his left breast, and a third with a mallet on his forehead). The murderers hid the lifeless body at night in a forest, planting an acacia branch on its grave. Hence the birth of the main Freemason symbols and that almost all sciences are plagiarized with elements of the Kabbalah (a study that understands the meaning and weight of numbers and letters) and silently infused by the so-called "Zionists".

THE TOP OF THE PYRAMID

Related to this we see a notorious Illuminati-Freemason symbol: the pyramid. This logo identifies the hierarchy and is a Hebrew symbol given to David, in which the institution or order of government in the universe is seen. The pyramid seen on the dollar bill and many corporate emblems is the clear example of who is honored. The apex of the pyramid separated as at Cheops - from which the apex disappeared - symbolizes that the head of the main corner of the pyramid or "the king of universal rule" will be removed or prevented from taking his place. For this reason there is an eye of Ra (a word that in Hebrew means "to see") that observes and records everything, enslaves humanity and replaces the true king. But, who is this "king" whose insignia represents the top of the pyramid? The answer is in the New Testament: «*Jesus said to them: Have you never read in the Scriptures: The stone that the builders-masons rejected, Has become the head of the corner. The Lord has done this, and is it a wonderful thing in our eyes? Therefore I tell you, the kingdom of God will be taken away from you, and it will be given to people who produce the fruits of it. And whoever falls on this stone will be broken; and on whomever it falls, it will crush him.* » (Matthew Levi 21:42-44, The New Testament)

Another sense of Freemasonry is deity. Officially it is said that they are irreligious, but the main teachers and prophets of

Freemasonry venerate "Iblis", which is another name for Satan, but only those promoted beyond the 30th degree are aware of this truth, and therefore from this level. From now on they are considered "enlightened" (Latin: "Inluminatum"), because now they do know who they really serve. [*Data under revision of the 20th of August of 2023: the word 'Illuminati' don´t seem to come from Latin, but from Sumerian, due to the gods 'Ilu', for the monarchies consider them selfe descendent of the Mephilim.*]

Albert Pike – A 33rd degree Freemason, founder of the Ku Klux Klan. Freemason, Sovereign Grand Commander of the Eastern Jurisdiction of the Ancient and Accepted Scottish Rite. Great Sovereign Commander of the Supreme Council of Great Sovereign Inspectors General of the 33rd degree. Magnificent Commander of Freemasonry. Confederate general in the American Civil War who was convicted of treason and imprisoned. Albert Pike said: "*Freemasonry, like all religions, all mysteries, conceals its secrets from all but the adepts and sages, or the elect, and uses false explanations and smug interpretations of these symbols to deceive only the deserving.*" *be deceived For you, sovereign great inspectors general, we say this, that you can repeat it to the brothers of the 32nd, 31st and 30th degrees. The Masonic Religion must be maintained by all of us, the initiates of the highest degrees, in the purity of the Luciferian doctrines. If Lucifer were not God, could Adonai and his preachers slander him? Yes, Lucifer is God, and unfortunately Adonai is also God. Baphomet, the hermaphrodite ram of Mendes, is the vital principle to which worship has historically been rendered.*"

Giuseppe Mazzini: 33rd degree Freemason, founder of the Italian Mafia. He fought to establish the republic in Italy, expelling Pio Nono from Rome, the Pope who excommunicated the Freemasons. In 1834 he assumed the leadership of the Illuminati lodge, holding this position until his death in 1872.

Mazzini said: «*We must let all the federations remain the same, with their mechanisms, central authorities and different modes of correspondence between the high degrees of the same rite, organized as they are at present; but we must create a "super rite" that remains unknown, for which we will only summon those masons of a higher degree whom we select. With respect to our Masonic brothers, they must swear to keep their activities under the strictest secrecy. By means of this superior rite we will control all the Freemasons giving rise to a single international center, the most powerful because its direction would be unknown. The members of the first three degrees of Apprentice, Companion and Master, are ignorant of the meaning of the symbols, and therefore are not yet in a position to have a physical or sensitive relationship with Satan, nor do they suspect that they can achieve it, since many of them they manifest that they do not believe in the Devil. But from the moral and intellectual point of view, however, they have a perfect relationship with Satanism: "whether they like it or not they are in our ranks."*»

Eliphas Levi: Renegade priest whose mother committed suicide because of him, became a 33rd degree Freemason, occultist, and black magician. It reached levels of sorcery not understood. He was the creator of the figure of Baphomet or the goat of Mendes (Egypt). And symbols like the inverted pentacle. This Jew said: " *Faith is superstition and madness. Which is more absurd and more impious than attributing the name of Lucifer to the Devil, Lucifer is the spirit, he is the paraclete, he is the holy spirit, at the same time as physical Lucifer the great agent of universal magnetism.* "

Manly Palmer Hall: Rosicrucian, Masonic author, and founder of the Society for Philosophical Research—which Erasmus Darwin fervently supported. He said: " *I hereby promise the great spirit Lucifuge, Prince of demons, that each year I will bring to him a human soul, to do with it as he pleases, and in exchange, Lucifuge promises to bestow on me the treasures of the Earth, and will satisfy my every desire* *throughout my natural life. And if I fail to bring him each year the offering specified above, then my own soul will be confiscated from him."* Signed.

Helena Petrovna Blavatsky – was the most important priestess of Eastern theosophy. Called "Madamme Blavatsky" she was followed and admired by Papus, Levi and many other occultists. Creator of the most satanist secret society in history. He said: « *Lucifer is the logos...the serpent, the sage. It is Satan who is the god of our planet and the only god. The heavenly Virgin who has been the Mother of the gods and demons at one and the same time, because she is the always loving benefactor deity... but in ancient times and in reality Lucifer or Luciferius is her name. Lucifer is the divine and earthly light, "the holy spirit" and "Satan" at one and the same time.* »

Count Alessandro di Cagliostro: false noble title with which Giuseppe Balsamo, doctor, charlatan, alchemist, occultist and High Freemason of Sicilian origin, toured the European courts of the 18th century. His most famous crime was defrauding a man out of all his money, claiming he had an aptitude for alchemy. Cagliostro claimed to have been born into a Christian family of noble birth, but was abandoned shortly after birth on the island of Malta. He also claimed that as a child he traveled to Medina, Mecca and Cairo, and upon returning to Malta, he was initiated into the Sovereign Military Order of the Warriors of Malta, where he studied alchemy, Kabbalah and magic. He founded the Egyptian Rite of Freemasonry in The Hague, and was influential in the founding of the Masonic Rite of Misraim (Hebrew: "Egypt", also the name of the chief son of Ham, the cursed son of Noah. His offspring plagued the lands of Canaan with the occult, satanism and witchcraft).

His Masonic name was Ouroboros [*which is the name of the serpent that bites its tail, and which identifies the worst of all hells*]: Due to his many crimes, he was sentenced to life imprisonment and ended his days in the dungeons of the castle of San León, in

Urbino in the year 1795. Cagliosotro was the one who instituted the use of the triangular altar, the Shekinah, in the center of the Rosicrucian Temples. He claimed to be possessed by powerful spirits that granted him the impressive power to work miracles even before the most critical and incredulous. It made occultism fashionable and popularized esoteric and magical practices in social environments that were naturally averse to these black practices. He said: " *Evil and sin are states of mind, there is the sanctity of sin.*"

Papus: was a 33rd degree Freemason, Bishop of the Gnostic Church of France, and a member of the Ahathoor Temple of the Golden Dawn, who became a member in October 1887 of the French branch of H.P. Blavatsky's Theosophical Society in the Isis Lodge. Creator of the Order of Unknown Superiors, commonly known as the Order of Martinists. Organized the international Masonic conference. It was precisely in the course of this conference that Papus received from the Freemason Theodor Reuss the patent to establish a Supreme Grand General Council of the unified rites of ancient and primitive Freemasonry, and, quite possibly, control of the OTO (Ordo Templi Orientis). He said: "*When the initiate completes the initiation and meets Baphomet, also called Lucifer, it can be said that he has discovered his real power.*"

SL Mac Gregor Mathers: Luciferian and 33rd degree Freemason. Founders of the Golden Dawn, Praemonstrator of the Golden Dawn Temple of Thoth-Hermes. They said: « *Perhaps for brevity we could say that the evil in the universe could be understood as unbalanced energy, or applied in the wrong place and that by rescuing the dark parts of its being, which are in relation to the demonic hierarchies. The magician is submitting them to his true will to become a whole being freed from the contradictions that coexist within each one.*»

Aleister Crowley: 33rd degree Freemason, he founded the Astrum Argentum, the order that was to supplant the Golden Dawn and extend Thelema. Head of the English OTO since 1912 and its international manager since 1922, he spread Thelema and made it known to the world. Member of the Golden Dawn, he was the greatest magician and occultist of the 20th century. He said: «*I will build a new heaven, I just went close to Satan and I still don't know why, but I found myself passionate to serve my new master. To practice black magic, one must violate every principle of science, decency and intelligence.*"

Adam Weishaupt – Renegade Jesuit priest, Freemason and occultist, creator of the Illuminati, father of the plan to eliminate governments and religion, to establish the New World Order. He said that the monarchy, any type of organized government, private property, inheritance, patriotism and the family should be abolished. He was the one who ordered the infiltration of Freemasonry to make it the base of his movement.

ORGANIZATIONS CREATED

David Icke, one of the great researchers on these lodge-sects wrote about it: « *The men who go to their local lodge in your city will not have the remotest idea of how their organization uses them. For the plan to work, you have to keep them in the dark, and what better way to do that than through the various levels of initiation. Only the ones they call "acceptable" progress to the higher levels and find out what is really going on. The vast majority of Freemasons occupy the lower three levels. They are the cannon fodder of the organization. Between grades 4 and 33, you will find those who, according to them, "think straight", and who have influence in society up to the presidents of the United States. After the 33rd degree there are the "Illuminati degrees". Something that is not mentioned in any Freemasonry manual. The latter are the ones who control the show and are agents of the "All Seeing*

Eye" sect, the engines of the New World Order. Global Freemasonry is a huge pyramid of manipulation that controls the entire planet."

One of the satanists of this caliber was, for example, the former president of the USA and 33rd degree Freemason, Franklin Delano Roosevelt, and also to say that practically all the presidents of the USA – except 3 of them, among whom were Abraham Lincoln and John F. Kennedy - were Freemasons and members of secret societies, for example, George W. Bush, both the son and the father, belonged to the Skull & Bones. Bush Sr. himself was director of the CIA when it assassinated John F. Kennedy and is an important part of the core of power and control of MJ 12.

[Data added in the review of this book on 08/19/2023: many suspect that George W. Bush Jr. was behind the death of John F. Kennedy's son, who would have been his main competition in the 2001 presidential election, of not having died in strange conditions while flying over the Gulf of Mexico; and the work by Milton William Cooper (Naval Intelligence officer), 'Behold A Pale Horse', argues that Bush was hired by the elite in the 1980s to direct drug trafficking from Colombia to the US. that the Bush family provided much of the weapons to both sides of the conflict in World War II.]

Now we can understand the potential interest of Darwinism in establishing evolutionism as a truth, despite not being subject to real scientific support. Could Freemasonry have so much power to give birth to the new naturalistic philosophy? There is a strong influence of the Masonic lodges, both today and throughout the history of humanity, with destructive and enslaving interests. The Freemasons-Illuminati have created and strongly influenced, among many others, corporations, organizations and global companies, among them we find briefly:

• **Religions**: Protestantism, Jehovah's Witnesses, "The Church of Jesus Christ of Latter-day Saints" (Mormons), Catholicism from the first century onwards, Orthodox, Anglicans, Jesuits, Jacobites and Calvinism.

• **Best-known universities**: Yale and Harvard.

• **Sects**: Raelians, Church of Scientology, Ku Klux Klan (KKK), Italian Mafia, Shiites (Ismaili order), Hashishim (Latin: "Murderers." Ismaili order).

• **Secret Societies**: Skull & Bones, Rolls & Keys, Vril, Thule, Jason Scholars, Hospitaller Knights, Teutonic Knights, Knights Templar and Rosicrucians.

• **Movements potentially promoted**: International Drug Trafficking, French and American Revolution, Sexual Revolution, Concentration Camps, Climate Change, Feminine Independence -so that more people are constantly working for the system-, Social Manipulation, blackmail with geomagnetic weapons, Corporatocracy (indivisible power of corporations), Independence, World War I and II, Persian Gulf War I and II, Afghanistan War, Vietnam War, support in the Holy Inquisition, Television Mind Control, Codex Alimentarius (with the WFP, for contaminate food and water along with the "Chemtrail" phenomenon), The Great Depression, the 1955 Crisis, the current Economic Crisis (2008 onwards), the Education System, Fashions, Globalization, the 9/11 Conspiracy, 11-M and 7-J, the Nazi Holocaust, Revolutionary Movements (terrorists), Inflation, Paper-Money System, Theory of Evolution and the "Blue Beam" project.

• **Philosophies**: Aristocracy, Nazism, Monarchy-Royalty, Revolution, Communism, Fascism, Anti-Semitism, Racism, Xenophobia, Machiavellianism, Anarchism, Monetary Aristocracy, Naturalism, Marxism, Anti-Christ, Dialectic and Martial Law.

Organizations: Pharmaceutical Corporations, Movie Companies, Banking Community, Monopoly Oil Companies, Continental Army, Music Industry, Star Wars (Space Militarization), United Nations (UN), NATO, Council on Foreign Relations

(CFR), Trilateral Commission (TC), Central Intelligence Agency (CIA), National Security Agency (NSA), WHO (World Health Organization), Majestic 12 (MJ12), European Union (EU), Asian Union, US Senate. US, US Congress, European Economic Area (EEA), US House of Representatives, the Round Table, European Parliament, IMF (International Monetary Fund), World Bank, Rotary Club, NASA, Secret Space Program, G-8, G-12, G-20, the Bilderberg Club (where all the world's leading media leaders participate) and the New International Order.

Mass Media

The owners of the most prestigious magazines, press and television channels are regular members of a modern international secret society: the Bilderberg Club. Because of this, the media continue to promote and propagandize the hoax. « *In conclusion, I would like to say the following: the greatest Masonic and humanist duty is to adhere to positive science and reasoning, to embrace the idea that positive science is the best and only way to [study the theory of] evolution, to spread this belief among the people and to educate the peoples in the light of it.*» Dr. Selami Insindag, Dean of Freemasonry. Some magazines such as "Scientific American", "Nature", "Focus" and "National Geographic", are publications on biological issues and respected natural sciences in the West, which make the theory of evolution their official theory and propagate it as something proven.

The information we have provided in what we have written so far strongly indicates that the theory of evolution has almost become the official ideology of the secular world order. And each and every one of the ideologies of this one - whether of the right or of the left - are based directly on evolution. This is why the theory of evolution spreads intensely throughout the world and is imposed on society despite the contradictions and absurdities that characterize it. In other words, those who support the unfounded theory of evolution and promote it as proven, are the "social forces" that consider its existence essential in order to survive politically and socially. The term "social forces" includes in its meaning a wide variety of groups, such as fascists, Marxists, capitalists, and racists. In any case, among these groups there is a particular "central force" that recognizes more than any other the socio-political importance of the theory of evolution and therefore conceives its popularity not only as a necessity but as "the duty larger".

This "central force" was totally involved in the construction of the New Secular Order and maintains the continuity of its derivatives, that is, of Marxism, fascism, etc. In addition, it has a disciplined and highly hierarchical internal structure. Consequently, it possesses the necessary apparatus to maintain social control. Of course, we are talking about Freemasonry, the greatest line of power in recent centuries, which established the New Secular Order and which, due to the war it launched against religion, began to idolize itself, receiving considerable support. of the main classes of the most "outstanding" countries of the world. Undoubtedly, the revival of religion throughout the world - in the words of the French writer Gilles Keppel, author of "The Revenge of God" - is the beginning of the collapse of Freemasonry. Even so, this situation does not make her passive. On the contrary, it makes her more aggressive: she is determined to hold the lead in the long war against God, both philosophically and politically.

Today's publications of the Masonic lodges are full of texts about the history of the fight against religion and always have the tone of the inescapable determination to continue with that war as is evident in the movie "Angels and Demons". (The Illuminati against the Catholic Church). For example, in one of the texts of the Masonic Lodge of Turkey entitled "What can be achieved today with the Masonic Lodges", the following message is given: « *Currently we have brothers at all levels of management and covering important obligations. All of them are responsible for fulfilling their duties within the structure of Masonic values. When we think about the dogmas that surround us and see the attacks on our institutions, even by people who were educated in them, we should believe that there are no limits to our objectives. If we do not engage fully in the political and social fields with the power that we have, we would not be proceeding appropriately with respect to our glorious history. Freemasons today must be determined to compromise with their brethren who came before them in serving Freemasonry.*"

When we analyze the media in the West today, we frequently come across news that is based on Darwin's theory of evolution. Popular, prominent, and "respected" newspapers and various media outlets continually promote this theory. The style used always carries the tone of an absolutely proven theory that leaves no room for discussion. The catchphrase most often used is " *the 'missing link' in the chain of evolution has finally been discovered*." In those cases and especially, a skull "discovered" in some remote place becomes important evidence for the theory that humans have their ancestors in monkeys. « *People who do not know anything about the theory of evolution see such news as absolutely true laws. The theory that living organisms came into existence by chance is accepted as law by all those who believe in science*." (Harun Yahya, The Deception of Evolutionism)

Those news are broadcast by the giants among the media through local newspapers in almost every country in the world. They deliver the news about it with a classic quirk that is common to almost all news outlets: " *According to Time, a major fossil discovery fills the gap in the chain of evolution*," or " *According to Newsweek, scientists elucidated about the unexplained points of evolution*." Those sentences always become headlines. It is interesting to note that all of them aim to make society accept the process of evolution as a proven fact. However, what is presented as proven facts and evidence are the falsified evidence that I will talk about shortly.

Apart from the media, the same picture is seen in scientific research, in encyclopedias, and in books on biology. Almost all of these sources present evolution as an absolute reality. In short, the media, which is under the control of secular and academic forces, fully upholds an evolutionary point of view and imposes it on the public. This influence is so strong that, over time, the theory became taboo. The denial of the theory of evolution is presented as something that contradicts science and rejects concrete realities. For this reason, especially since the 1950s, despite all that has been revealed, explained and openly proclaimed about the errors that underpin the theory of evolution, the press organs and scientific publications do not express a single critical word.

Douglas Dewar, a well-known evolutionist, in his detailed research on the birds of India, concluded that species cannot transform into one another and also explains the strong relationship between the theory of evolution and the media: "Only a *few few people could conceive why it is important for evolutionists to control the media. You don't see many newspaper articles today that oppose the theory of evolution. On the other hand, most religious newspapers are controlled by modernists who accept the fable of 'monkey ancestors'... Generally speaking, the editors of all newspapers regard the theory of evolution as a proven fact and those who reject it, they are called ignorant and crazy. Newspapers are edited by evolutionists who avoid publishing any article that even minimally criticizes the theory of evolution... The editors do not publish any book that opposes a popular theory since it would provoke objections from different sectors of society.*"

And even if the writer is willing to pay for the publication of the case, the publishing company would not follow through, understanding that it would lose credibility in public opinion. Therefore, society is informed only from one point of view and people then assume that the theory of evolution is a proven fact. Not surprisingly, since "facts" have never been important to the theory of evolution. The important thing is to "give life" to the Darwinian legend, despite the scrutiny to which it has been subjected. The propaganda disseminated by the media is not just limited to newspapers. Television channels and radio stations also make an important contribution to the propaganda of evolution.

Blue Beam

The masters of this world, unifying masonic lodges and objectives have interwoven all levels of society. They gather annually the most outstanding people in the world with the name of "Club Bilderberg" who have spent years creating and developing all kinds of plans to generate wars and globalize the monetary system and implant human beings with a biochip with which they know where they are going. and what they do, while humans with this device are vulnerable to being mentally manipulated. It sounds like science fiction, but in the present, with the technological advances that we possess, one can no longer speak of "fiction", but only of "science". These men are basically divided into two groups: MJ12 (the US Illuminati), an order of Freemasons that determine who is to preside over the National Security agencies, the Presidency of the Government and the Policy of the Nation; The second group are the European lodges, where royalty, politicians, corporation owners and other influential groups based on Freemasonry are.

The institutions that the Freemasons have created, such as the United Nations Organization (UNO), the Central Intelligence Agency (CIA) or the National Agency for Space Aeronautics (NASA) carefully fulfill their objectives. It is naive to believe that the world works as we believe that we see it on television: that is just media propaganda. The reality is that not even the cinema is spared from the influence of Freemasonry. So even Hollywood companies receive support in exchange for influencing the masses through the messages that are sent to us in each feature film. Is there any relationship between the Blue Beam Project and neo-Darwinian goals? Evolutionary theory is being bombarded through the movies more than any other medium.

Some of the many famous movies that encourage the imagination and allow for a favorable mutation or the possibility of an organism evolving are, for example: X-Men, Monsters vs. Aliens,

Godzilla, Dragon Ball, Underworld, Planet of the Apes, Hellboy, The League of Extraordinary Gentlemen, A Chance of Meatballs, The Time Machine, Water World... To this we have to include the number of films about fantasy, magic, and all kinds of impossible things that allow the subconscious and unconscious to conceive the feasibility of this type of thing happening, even though the person is consciously aware that it is only the big screen.

What happens with this is that when really supernatural or amazing things happen, they are either not so shocking, or they seem like part of a conspiracy theory. In this regard, we have the revelation of the Canadian Serge Monast about the Blue Beam project, after whose revelation he was assassinated. In it, he revealed a plan in several phases motivated by MJ12 and the Illuminati, and established in the UN to be carried out by NASA and the Armed Forces:

A. Overturn all archaeological, historical and educational knowledge, including the increasingly strict implementation of the theory of evolution in all areas.
B. Put on a whole holographic space show projecting a new god that the MJ12 have perfected.
C. Reception at low frequencies of "supposed" divine communications, launched from satellites.
D. Supernatural manifestations:

1st. Invent an alien invasion. This is why they promote so many movies about UFOs and ETs. The goal of this is to disarm all nations in an attempt to defend the planet from the fake invasion.

2nd. Deceive Christian believers with their long-awaited Rapture (return of Jesus Christ to remove his chosen ones from Earth. Ref.: Luke 17:34, Matthew 24:31 and Revelation 12:5).

3rd. Low Frequency waves used to manifest specters, poltergeists and satanic beings around the globe and make the world ready to receive its "new messiah".

It seems like a topic from some lunatic, but a former satanist, brother Jesús Salcedo, also a former priest of María Lioza, wrote in his book entitled "From a priest of the devil to a minister of Jesus Christ" that in a meeting on the Sorte mountain, where he was, one of the items on the agenda of these sects was that the sorcerers and demons are already prepared to torture and kill the evangelicals who remain in the period of the Great Tribulation. What do these things come out of? The reader is in his total right to accept or reject these affirmations. However, when all this takes place there will be no room to look back and try to change something.

SUCCESS

We certainly do not have to expect Freemasonry to carry out the "brainwashing" process openly. This practice goes absolutely against the traditional methods of the organization. A Freemason from the Turkish Lodge, Halil Mülküs, explains the method: «*Freemasonry does not execute anything itself. It shows the way to individuals. Those who share Masonic views and Masons who contribute to the development of new ideas, practice their professions at various levels. They exert influence in the universities, they are rectors, professors, statesmen, ministers, doctors, lawyers, etc. They impose Masonic views on society through the institutions where they are present.*"

It is clearly a question of who gets more propaganda and control over the masses: «*Before discussing the role of the media as a tool for the indoctrination of evolutionary views, it will be useful to explain the general function of the media. Under the slogan "news = life", the media claim to reflect what happens in real life. According to this, the media become our eyes and ears that perceive the outside world. Therefore, just as we trust our auditory and visual sense, we should trust the media. However, the reality is different. The main function of the media is not to communicate what actually happens in the real world but to format our minds with respect to the outside world in a certain way. "News = life" is true in a sense. We are required to learn about the outside world through what is presented to us as "news ."* (Harun Yahya, The Deception of Evolutionism)

The well-known North American linguist and scientist Noam Chomsky made the most detailed study about the role of the media in his book "Necessary Illusions: The Control of Ideas in Democratic Societies." In this work, he reveals how the media become a tool to control the social processes of ideas. In the first place, says Noam Chomsky, that in his country democracy is a totally different system from what we define as "democracy". According to him it is a hidden system and a type of invisible totalitarianism, since the system works, obviously, based on public "consent", although said "consent" is formed as a result of the process of "brainwashing" generated through some "instruments". Chomsky provides notable examples of invisible totalitarianism, which can also be called "democratic totalitarianism."

The examples he gives indicate that US policy makers use the media efficiently. In other words, when they decide to intervene abroad, they first take advantage of their irresistible magic to prepare public opinion. First of all, the US leaders present to the public as "demons" the targets they want to attack, such as Saddam Hussein, Noriega, Islamic groups, the Sandinistas, Bin Laden, etc. For this they efficiently use different methods of propaganda or psychological techniques. Consequently, the public applauds the invasion of a foreign country by US soldiers and gives its consent to the policies formulated by the different administrations, although in reality that consent is established by the political apparatuses. That's why Noam Chomsky defines this sophisticated totalitarian mechanism as "consent elaboration."

One of the most formidable examples of this method took place during the administration of former US President Woodrow Wilson. This example, considered by Chomsky to be " *the first modern government propaganda operation*", can be sketched out as a plan to convince the people to give consent for the country to go to war in the first world conflagration. During the first years of it, most Americans were determined not to participate. However, the centers of power, which had profound influence over the government, were interested in intervention in the armed conflict. Therefore, a commission was formed, called the Creel Commission, which took over the propaganda on behalf of the government. The Creel Commission managed to transform that passive people into another with hysterical characteristics in just six months with a strong will to destroy the German nation, go to war and save the world. As a product of that program America went to war.

A prominent theoretician of this totalitarian technique is Walter Lippmann, one of the best-known American columnists. He is one of the founders of the Council on Foreign Relations, an important unofficial institution officially concerned with US foreign policy - although in reality it is a functional apparatus of MJ12's relations with and control over other countries. This gentleman made every effort to develop the best control systems for societies through the elites and without anything getting in the way of the task. That is why Chomsky considers Lippmann "*the architect of the 'consent elaboration' theory to get the people to approve even unwanted decisions under the influence of new propaganda techniques.*" Lippmann argues that the government of a state should be run only by " *a special group of intelligent people who are capable of assuming responsibility, while the mass population should be kept entirely out of the decision-making mechanisms."*

According to Lippmann, the people are nothing more than "*a stupid herd*" and should not participate in the (government) administration process but have to remain obedient followers of the decisions. Chomsky emphasizes that Lippmann's totalitarian approach is very similar to Leninist theory. Current US policy represents a system based on what Lippmann refers to as "consent processing." Undoubtedly, this situation points to a reality about the current democracies represented by the US and Western countries: in these countries "sovereignty" is not in the hands of their respective peoples but is captured, evidently, by the power that controls the process of ideas at a massive level as happens in Venezuela with Hugo Chávez and the media, for example. In this context, the media is used as one of the most important tools to control the thought process.

Of course, not all media can be put under this category. However, "the giants among the media", present in almost every country in the world today, fall into that category of "controlling tools". This is why, in some cases, despite the alleged open opposition to governments, the media have intimate relations with the powers that be in charge of "governments" being themselves part of the Bilderberg Group, or "the Masters of the World". In any case, the "control over ideas" is not limited only to political issues since the power that is established today is still partly religious, and therefore the anti-religious perspective should be encouraged, that is, the abolition of authority. religious. That power can continue to be maintained only if anti-religious views continue to be widely accepted by society. The acceptance of religious views and viewing them as the only legitimate source of conduct is totally unacceptable to the established system.

SPONSORS

Despite the contradictory nature of Darwin's so-called theory, it was widely adopted as it provided a kind of explanation that served to fill the great hole of materialism and the secular order in its broadest sense. A group of scientists voluntarily took over the task of promoting this theory. The best known of these was Thomas Huxley, who was known by the nickname "Darwin's bulldog." Thomas Huxley, whose ardent defense of Darwinism alone was responsible for its rapid acceptance, brought the evolutionary hypothesis to worldwide attention through the well-known "Oxford discussion," that is, the discussion he held at 1860 with the Bishop of Oxford, Samuel Wilberforce.

It is not difficult to understand why Huxley dedicated all his efforts to propagate such a theory of evolution if we take into account his "societal ties". Huxley was Dean of Freemasonry and, like other participants in it, a member of the Royal Society, one of the most important scientific institutions in England. All of them explicitly and in detail promoted the alternative theory of natural selection foreshadowed by Erasmus Darwin. This Masonic institution gave so much importance to Charles Darwin and Darwinism that some time later it began to annually award successful scientists with the "Darwin medal", in the style of current Novel prizes. In other words, it was not only Darwin who carried out this mission. Freemasonry, one of the most important headquarters of the war waged against religion, provided complete support for this hypothesis the day it was presented.

The evolutionary hypothesis, although it did not convince many people when it was first defended, gained immense popularity in a few decades due to the ideological support it received. And it is because of that support that Darwin's followers were unmoved when biological studies disproving Darwinism were presented. Did we come from the monkey? If we see it from the doctrine of evolution, the answer is a certain "NO".

But what about the fossils? We have been led to believe in the last two centuries that there is an overwhelming body of evidence supporting the hypothesis that modern man arose from a process of evolution of "lower" organisms that began in the sea with the first cells. Said ideology is clearly a way of wanting to see life, but faced with the weight of reality, it is nothing more than a fable that has never been observed through a microscope and recorded on a camera. And what are the physical evidences that evolutionism supports to say that we come from the chimpanzee? Why only man evolved? Why didn't other species continue to evolve? Monkeys are still monkeys; the fish, fish; reptiles, reptiles; the birds, birds; and mammals, mammals.

None of these species has evolved and even hundreds of species have become extinct because they have not been able to "adapt" to their environment. The information collected throughout the world shows that before the peculiar hominids and Adam (Adam, from the Hebrew "the Most High in the blood") and Java (Eve), the Earth as well as the Universe already existed and are " populated", that is why Jesus himself emphasized: *"In my Father's house there are many mansions; if it were not so, I would have told you; I am going to prepare a place for you."* (John Zebedee 14:2)

NEW WORLD ORDER

According to a study by David del Fresno, today we see that within the most forceful dogmas of faith of our current society in reference to knowledge, the first commandment is, strange as it may seem, the conception of human development as the potential enemy of all animals, men, plants, and in general, of the entire planet. According to him, this is not something new: «*It is an intellectual process that has its roots in the dawn of the 20th century: In a process very similar to what is happening now with the controversial use of the so-called "stem cells", for that then the conception of man as a mere animal, stripped of the dignity that is intrinsic to him, and that places him above the rest of the animals began to circulate in the western intellectual, scientific, academic and political circles.»* As surprising as it may seem to us, famous people who supported these ideas included:

- Alexander Graham Bell (inventor of the telephone)
- Margaret Sanger (founder of the abortion organization *Planned Parenthood* (which translates as: "Planned Parenthood"), [*directed by the Bill Gates family, and he being its acting director)*], as well as several Nobel Prize winners.
- George Bernard Shaw (famous playwright)
- Lelan Stanford (founder of the university that bears his name)
- HG Wells (novelist and author of "The War of the Worlds")

According to David del Fresno, several states in the US were at that time passing various laws that promoted eugenics (the elimination, through forced sterilization, of "inferior" races). These legislative efforts would not have been able to prosper without the explicit support of three important North American academic and scientific entities:

- The National Academy of Sciences
- The American Medical Association
- The National Research Council

Who financed and directed these scientific entities? Well, nothing more and nothing less than the Carnegie and Rockefeller foundations, preferably. Under the tutelage and direct promotion of these two foundations, eugenic theories were soon accepted in Europe and more specifically in Germany, being financed until 1939, a few months before the start of World War II. The German eugenicists came to progress so much that, from 1920, the world leadership of the pro-eugenics movement corresponded entirely to the Government of Germany.

After the supposed Nazi defeat - since the Russian and North American secret services, together with the high commands of said governments knew that Hitler and the most prominent of the German army had fled to New Swabia (South Pole) -, the eugenicists disappeared, but only apparently: In reality they had only changed their name, beginning to call themselves "Social Darwinists" in honor and memory of Charles Darwin. We already see that Darwin had found an explanation that, according to him, proved the existence of an evolutionary process as a natural law, which he called "survival of the fittest". However, the Darwinian explanation was based on a logical fallacy called circular reasoning: *"Who are the fittest?" Those who survive. And who are the ones who survive? The fittest* ." Despite this obvious fallacy, the Darwinian principle soon generated a whole ideological structure that came out of the blue to the Rockefeller clan, which found in the Darwinian formula of the survival of the fittest the moral justification it needed to carry out its activities aimed at eliminating competition. The Rockefellers and their international banking associates pooled their power and fortunes for the purpose of promoting Darwinism and other similar ideologies, and to this end they created two organizations under their umbrella that would serve as their command center to coordinate their efforts in America and Europe:

- The CFR, or Council of Foreign Affairs (New York), which in Spanish is 'Consejo de Relaciones Exteriores'.
- The Royal Institute for International Affairs (London), or

Royal Institute of International Affairs.

Both organizations apparently integrated into another older structure called "The Round Table Group" (The Round Table Group) of Masonic tradition according to some famous analysts, and branches of the cover that bears the name of "Council of Foreign Relations" (CFR - Council of Foreign Relations). As unusual as it may seem, the philanthropic organizations that depend on the Rockefeller clan are based on the same purpose: A cursory study of where they channel their funds shows that most of their efforts are dedicated to financing third-party organizations whose purpose, both direct as indirect or covert, is the control of population growth.

Not surprisingly, in a study published by the Club of Rome (one of the many organizations linked to the Rockefellers) it is explicitly stated: « *If the fight against a new enemy unites us, we have conceived the idea that pollution, global warming global economy, water scarcity and hunger are the perfect enemy. And all those dangers are caused by human intervention. Therefore, the real enemy is humanity itself*." This is not correct, since the fault lies with the estates that they themselves have established. The human being is nothing more than a puppet that has been discreetly directed towards the situation in which the planet finds itself today.

To carry out their plan to reduce the world's population, affirms David del Fresno, the Rockefellers and their partners -among which are the Morgan, Rothschild, Whitney, Bush and royal houses- start from this main idea: « *Better Rather than make use of a violent process that may seem unacceptable and generate an uncontrollable rebellion, it is better to make use of a gradual process, presenting it in such an attractive way that makes it acceptable to the majority of the population.* " With this in mind we can see that its objectives are clear:

- 1° The drastic reduction of the population. The great thing is to eliminate 93% of the global population.

- 2° The reduction of consumption to pre-industrial levels.

It becomes evident why all the great themes promoted by the Rockefellers since 1960 lead in one way or another to achieve this result. However, after devoting their efforts and their money to this end for many decades, the Rockefellers and their associates have come to the conclusion that the growth of the planet's population is uncontrollable, unless drastic measures are taken to stop it. The best card is to cause a Third World War, and the best candidates to be the "spark" are Jews and Muslims. And the only way to achieve all this agenda is with the implementation of a global political and economic system that they have agreed to call "New World Order" (New World Order). David del Fresno adds that its promoters start from the following premises:

- 1° The democratic political system is exhausted.
- 2° The capitalist economic system is exhausted.
- 3° Christianity is the biggest obstacle to the implantation of a new and necessary "universal ecological mentality".

« This is why more and more symptoms are being perceived that there is an intellectual process designed to destroy Christianity and replace it with a new universal ethical system that consecrates ecology as the new universal religion. Is it then understood why the rise of Gnosticism and the beliefs of the Freemasons, as well as the success of novels, comics and cartoons in which the veneration of Mother Nature is exalted as the foundation of truth and good? All of this is part -in our opinion- of a process that is not accidental, and that we believe contributes to inoculate in the less-formed impressionable minds an acceptance, as favorable as it is inadvertent, of an entire system of beliefs and values fraught with relativism, environmentalism pantheism, occultism, magic... Of course, nothing that has to do with the promotion and defense of human life, the family, and the dignity of the human

being. Who can benefit from this whole process? It will be part of another analysis that we will reveal later.» (David del Fresno. November 19, 2009. CAMINEO.INFO)

Let us consider that once these men have achieved the objective of implanting a monetary system by means of a biothermal chip in the right hand of each human being, when they have also given all the military power of the Earth to a single organization (UN), and together with this establish a single world political power led by a single man who will establish a new and unique religion, then the world will be led to total slavery and spiritual collapse. This new system implemented by these members of the Bilderberg Club already has a name: "The Beast", and it has its headquarters in Brussels.

Knowing this, the Benjaminite missionary Paul wrote to the Greeks of Thessalonica: « *Let no one deceive you in any way; for he [Christ] will not come without the Distancing first coming, and the Man of Sin, the Son of Perdition, who opposes and rises up against everything that is called God or is the object of worship; so much so that he sits in the temple of God as God, posing as God."* (Paul's 2nd letter to the Thessalonians 2:3-4) Ergo, about this man, who will be the leader of the Earth for 3 and a half years, the Jewish missionary John also wrote on the Greek island of Patmos: «And he *made all, small and great, rich and poor, free and slaves, were to have a [subcutaneous] incision placed on their right hand, or on their forehead; and that no one could buy or sell, except the one who had the mark or the name of The Beast, or the number of his name.*» (Book of Revelation of John, "Revelation" 13:16-17)

The Jew Jesus of Nazareth himself mentioned this "estrangement" about which Paul spoke to the Thessalonians when they asked him about his return: «I tell you *that on that night there will be two in one bed; the one will be taken, and the other will be left. Two women will be grinding together; the one will be taken, and the other left. Two will be in the field; the one will be taken, and the other left."* (Doctor Luke 17:34-36) Dear reader, one leads to another. Everything is part of the gear.

EUGENICS

According to Sir Francis Galton, *"eugenics is the study of agencies under social control, which enhance or impair the racial qualities of succeeding generations physically and mentally. »* Darwin's cousin, Francis Galton, who is credited as the father of eugenics, saw an opportunity to catch up with the human species, taking the reins of Darwin's theory of evolution to apply it to social principles, of so that a social Darwinism developed. The Darwin, Galton, Huxley, and Wedgwood families were so obsessed with their new theory of social design that they swore their families would only procreate with each other, and they did. They wrongly predicted that, within just a few generations, they would produce supermen. The emerging pseudoscience only helped to strengthen the practice of inbreeding, which was already popular among the elite for millennia, the 4 families experiment was a disaster, in less than 2 generations of inbreeding, close to 90% of their offspring died at the time of inbreeding. born or had serious physical and mental handicap problems. Darwinism is the scientific appendage of Adam Smith's Free Market theory.

Darwin was based on Herbert Spencer, specifically from his book "Social statics", from where he got the phrase " *the survival of the fittest* ", which came out 5 or 6 years before Darwin's book on the "origin of man". Decades later, Sir Horace Darwin, son of Charles Darwin, was the 2nd president of the eugenics society, which prohibited those they considered imperfect from reproducing. In the US, eugenic laws came out then and eliminated thousands of people with the excuse that ills, drunkards, the poor and the like were hereditary matters, mainly blacks and Hispanics. Sir David Frederick Attenborough, the naturalist and Darwinian popularizer belongs to one of the great secret societies of powerful scientists, such as Optimum Population, who promote eugenics. Jacques-Yves Cousteau himself, the underwater explorer and photographer, was a eugenicist, but the banner of leadership for this movement has been the Rockefeller family.

Now there is another who leads a new group, which is Robert Edward Turner III, who is precisely an American magnate who is the founder of CNN, who says that 90% of the population is left over, so that we can live properly. Darwin said that the man was in an evolutionary process between the gorilla and the man, and that the woman was in the process of evolution from the primitive tribes towards the man, but below him. He also said that the workers and the poor, to control bad dispositions, are either executed or put in jail for a long time.

Adolf Hitler's own book "Main Kampf" (My Struggle) is based on Darwin's "Origin of Species", or "the maintenance of favored races in the struggle for existence". As another tycoon put it: " *The maintenance of a great business is simply the survival of the fittest...it is simply the combination of a law of Nature with a law of God."* (John D. Rockefeller, 1839-1927) It is like calling it "genetic determinism" that is, a law of justification, to do with the races what they want, or today, with the economy, what they want, because it is law of survival of the fittest. More recently, Richard Dawkins, a Darwinian zoologist, published the work "The Selfish Gene" in 1976, where the evolution of species is interpreted from the genetic point of view, saying that genes kept fighting and competing against each other. everywhere, and suddenly there were some who became superior and advanced the development of a species... something totally insane.

Thomas R. Malthus published in 1798, 'Essay on the Principle of Population', which is about the struggle for life. It is an outrageous book which discriminates against the poor just for wanting to add: population + consumption per person, and similar things also recently raised, and publicly, by Bill Gates (TED 2010 conference). In short, *"if you don't have someone to support you, you don't have the slightest opportunity for a ration of food, that is, those who don't have possibilities, may die."* (Criticism of Máximo Sandín, Bioanthropologist and professor of Human Evolution at the University of Madrid, against Darwinism)

Part II

DETERMINISM IN THE UNIVERSE

THE ALIENS

As the reader will be able to verify, our objective is not to promote Creationism, but Panspermia, with the substantial touch of the Holy Scriptures. Because? Because these were inspired by aliens. As can be verified by studying my other works, it is a fact that extraterrestrial civilizations exist and have been involved in our history. Let's make one thing clear, Moses did not invent the Tables of the Law or the Torah, but received it from superior civilizations from the cosmos. The ancient cultures called the astronauts "gods", and the "aliens" they called, under the Platonic and Socratic prism, "angels" and "demons". Although, most of the astronauts were, nevertheless, aliens, but they posed as gods. This action provoked the wrath of Jehovah, who allowed the Flood to eliminate this problem, and to the lineage that these aliens had with terrestrial humans. None of this is secret anymore, it is simply not something that is shouted to the four winds in the media or in the education systems, for mere vested interests.

Precisely, the famous book by the American journalist Michael Drosnin, entitled "The Secret Code of the Bible", is a clear example of how this advanced group of brothers from the galaxy gave Moses, and the prophets of Israel, a quantity of coded information, just as it happened with John when he received the future vision that we popularly call "Apocalypse" (which means in Greek: "Revelation"). In other words, the Bible is a universal record of knowledge codified in Hebrew —and even in Greek-, with numerology, symbology, terminology and mathematics. We talk about quantum in an apparent religion book. For what purpose? That the "connoisseurs" say with the keys of the origin, situation and destiny of the human race, and even of other 31 planets in our Galaxy. The Bible itself is not the only information in this regard, since there are dozens -and perhaps hundreds- of manuscripts that accompany this documentation.

Although disparagingly called "pseudo-epigraphical", mystically "apocryphal" and "deuterocanonical", all this material leads us to understand that we live in the midst of a war that has developed in our universe many thousands of years ago, and in the midst of which we are blindly. The reason for the coming of Christ, from the 6th - or in modern terms, of higher dimensions -, is crucial, since the destiny of each one of the human beings depends on it (I explain this in detail in my works "Reaching the Deity" and "A Forgotten Message", with certain relevant nuances in "Path to Mastery").

SCIENCE AND FAITH

Science, as the study of things, through 13 basic methods, whose peak is physics and chemistry, has not given rise, in the last 2 centuries, to the "spiritual" genre. The culprit for this has been the Catholic Church. Saying that they represent God, they have blinded the world with their deceptions, preventing humanity from accessing God. By focusing on the war against the Vatican, many have included God and Jesus in history, believing that they are guilty of the actions of Rome, or in any case, they are their inventions to control the people. The point is that practically all the concepts that, thanks to Catholicism, humanity has about God or about Jesus are false. Faith is not just believing, or believing without seeing. Faith is the conviction of something based on the visualization of it. If you don't understand what they're talking about, what do you believe? I say don't believe in George Bush. Does that mean that I don't believe he exists or rather that I don't believe in what he represents or presents to the world? To believe or not believe in him, for example, I must know what he is and what he is looking for. However, about God, obviously, it is difficult to believe because what is known about him is rubbish, because of the Roman Empire. From Catholicism came Protestantism, Anglicanism, Orthodoxy, Charismatics and many others.

Subsequently, from these, especially from the Protestants, came most of the popular trends, including Evangelicals, Adventists, Baptists, Calvinists, Lutherans, Pentecostals, Methodists, and Unitarians. Although all those who were weaned from Rome did so by contradicting it, they continued, in the long run, believing practically all of its nonsense and inconsistencies associated with what is mistakenly called "Faith". To understand what Faith is, we cannot go to modern Christianity —or to that of the last 14 centuries-, because, as such, it is contrary to the teachings of Jesus Christ. That's right, and it's ironic, because they bear his name.

Therefore, to understand this one must go to the source: the Jews. It is not about pro-Zionism, but about analyzing the source of things. Today there is so much misinformation about all these things that a whole school of learning is done to remove so much social poison and begin to see clearly.

Rome did not write the Bible, and let us begin by shedding light on this point, but it was Jerome who compiled all the literary material of the Israelites, which was available in the Roman Empire. In the past, the Bible did not exist, as such, but a number of Hebrew scrolls were converted into the ancient Greek language in Alexandria. The work of 70 Jewish scholars, called the Septuagint, compiled and translated into the language of the time - before Jesus - the entire history of the people of Israel. This writing included the Torah (the law of Moses, exposed in 5 books called Pentateuch in Greek, that is, "Five Blocks"), the Nebiim (the stories and warnings of the prophets of Israel) and the Ketubim (historical writings, basically of judges, kings and other important personages). This set of 3 groups is called TANAK (or Tanaj), which the West called Old Testament, because they associated it with the Covenant of Yahveh (Jehovah) with Abraham. All of this occurred before the birth of Yeshua (Jesus), possibly under Asmodean rule.

With the revolution that Jesus generated, the most powerful of the Sanhedrin in Jerusalem saw a danger in leaving the Septuagint in the hands of the people, because with it the followers of Jesus could demonstrate that he is indeed the Messiah who had been promised to Israel. For that reason they changed their strategy and removed some texts, transferring the study of the scriptures to the local language. What the Jewish religious leaders called Those of the Path or the Sect of the Nazarenes, that is, the Messianic Jews (primordial Christians), became a headache for them, and they began to harass them in all regions where Rome ruled and where there were Jewish communities and/or synagogues. The main stakeholders in suppressing the story of Jesus were those who held power in Jerusalem, and were important members of the Sanhedrin.

The miracles and the growth of the masses was such that the religious Jews who were against the Nazarenes or early Christians, took pains, until today, to eliminate all historical evidence of Yeshua (Jesus) among the Israelites. Centuries passed, the Roman Emperor Constantine saw his kingdom enveloped in a new religion and wanted to adopt it for himself and monopolize it. The followers of Jesus had already died, and Christianity had become Hellenized, mixing with pagan European traditions and beliefs. The next was to disrupt the history of Israel: they erased the name of Judea from the maps of Rome, they removed and hid books of the history of the Hebrews, they added phrases to the literature of the Israelites, they adopted Latin and Greek words to replace the Hebrew terms. relevant, they manipulated entire sentences, and to conclude, they prohibited the reading of this literary genre until a few decades ago. It is evident that, with all this manipulation, today people have such misconceptions about the archetype of God, the figure of Yeshua (Jesus) and about the Bible.

The letters and historical versions of Jesus were compiled by Rome and included in the New Testament genre, understanding that Christ made a New Covenant. The compendium of the Tanak and the material on Jesus and his apostles, in its context, received the Greek name "Bible" (Compendium of Books). However, Latin, and even Greek, notoriously blurred the "biblical" teachings, and it was precisely Latin that was the weapon of manipulation that was magnified with the translation of Jerome, called "Vulgate" (for the people). Rome, to define what they were opposed to, called them "heretics", but that word simply meant "of a different opinion". Where is the mystery? Those who wanted to distinguish in a humiliating way were called "pagans", which in Latin means "peasants"; Those who wanted to make gods before the people were called "saints", that is, from the Latin "santi", which translates: "consecrated".

They called the books that they wanted to leave out of the official order "Deuterocanonical", that is, "Measured, Set Apart", and those that they wanted to keep totally out of the access of those under religious control, they called "Apocrypha", which means in Greek: "Kept", "Secret" or "Hidden". Thus, with this mixture of Greek and Latin, they created a mystical and pejorative vocabulary. The term "spiritual" came from the Latin "subtle", but it was visualized as something etheric or ghostly, since in Greek it referred to "air" and in Hebrew to "wind". How things change from one language to another! The philosophical terms of the messengers of the Greek gods, "daimones" (demons), were adopted as malefic figures, and the other messengers, among the gods themselves, were the good ones: "aggelou" (angel). That is, the Greek, the word "herald", "messenger" or "emissary" is angel. There is no myth in this either. Saying "religion" is talking about "linking again" or "reuniting"; To speak of hell is to speak of a non-biblical legend, since it does not exist in the Holy Scriptures. So, with these examples, I hope the reader will visualize the disparate relationship between the original Biblical teachings and history and the amount of deception that arose when the "Bible" fell into the hands of Rome.

For this, and many other reasons - which you can better study in my book 'Reality or Religion?' -, supposed connoisseurs of symbology and ancient history have appeared who stumble in their analysis, since they start from their own Western cultures, and not from the Hebrew root, because they believe that Israel emerged from Egypt, the Bible from Rome and Jesus from mythological characters Hindus. For example, out of mere ignorance, people believe that Genesis refers to the "genes of Isis", since they start from Roman history, which certainly adopted Egyptian criteria, but not from the simple Greek language, in which language genesis simply translates:

"generations", because that is what this book is about: the generations of Israel, starting from Adam. Since the Creation tradition was practically the same everywhere, the Sumerians had their own version, almost identical to the account of the first chapter of Genesis, which they called "Enuma Elish".

On the other hand, this has already become the reason for multiple distortions, misinformation and debate, such as the term Zion (Zion, or Tzion), which refers in theology to the eschatological idea of the future Government of God on Earth, the which would be framed in Mount Hermon (current area of southern Lebanon). Zion, Tzion or Zion was an ancient reference to expect the Kingdom of God coming to our planet, but in the last centuries it was the name adopted by a financially powerful elite of Jews, whose interest was to establish, at their own expense, this kingdom, and to do it at the expense of the State of Israel, without the knowledge of ordinary Jews and Israelites.

CHRIST AND HIS COUNTERPARTS

DIONYSIO. After many centuries, all those who resented Rome, mostly Protestants or Freemasons -before Freemasonry and the like were introduced into the Vatican-, sought all kinds of arguments against the Catholic Church. This imperative desire led them to believe that they found in other characters, basically mythological, homologues of Jesus Christ, since they considered that Jesus was an invention of Rome to have control over the masses. Without taking into account that the main deity of Rome is the Virgin Mary, the restructuring of the Egyptian goddess Nefertiti, the figure of Jesus was taken as the target to crumble. For example, researcher Barry Powell believes that Christian notions of eating and drinking the "flesh" and "blood" of Jesus were influenced by the cult of Dionysus, or Dionysus.

Since wine was important to Dionysus, who was imagined as its creator, it was believed to find a parallel with Jesus, through the miracle of the Wedding in Cana: «On the third day there were *weddings in Cana in Galilee; and the mother of Jesus was there. And Jesus and his disciples were also invited to the wedding. And lacking the wine, the mother of Jesus said to him: They have no wine. Jesus said to her: What do you have with me, woman? My hour has not yet come. His mother said to those who served: Do everything I tell you. And there were six stone jars for water, according to the purification rite of the Jews, in each of which two or three pitchers could fit. Jesus said to them: Fill these jars with water. And they filled them to the top. Then he said to them: Take out now, and take it to the steward. And they took it away. When the steward tasted the water made wine, without knowing where it came from, although the servants who had drawn the water knew, he called the husband, and said to him: Every man first serves good wine, and when they have drunk a lot, then the lower one; but you have reserved the good wine until now. This beginning of signs did Jesus in Cana of Galilee, and manifested his glory; and his disciples believed in him.*" (Apostle John 2:1-11)

In the 19th century, Bultmann and others compared the two themes and assumed that the Dionysian theophany was transferred to Jesus. Heinz Noetzel disagrees, arguing that Dionysus never really turned water into wine. Martin Hengel retorted that opposing traditions would be anachronistic, and that since all Jews in the region were familiar with the turning of water into wine as a miracle, the Messiah was expected to perform it. Peter Wick argues that the use of wine symbolism in the Gospel of John, including the Wedding Wedding at Cana story in which Jesus turns water into wine, is intended to show Jesus as superior to Dionysus. Although, the discrepancies cannot be greater: Jesus has two main characteristics: He is the representative of the Heavenly Father and he is also the representative of the human race. Also, in addition, he is the king of the universe. However, Bacchus or Dionysus, was the god of wine, revelry and parties. He is one of the 12 of the assembly of Zeus.

It is clear that there is no parallelism, since even Zeus, his father, was a deity subject to an exclusive region (the heavens). Jesus himself, for example, was raised by his biological mother, who later had more children, while Bacchus was raised by nymphs, since his mother was charred, not on purpose, by Zeus. Another version says that his mother was the queen of the underworld, and this is indeed a radical antagonism. Jesus was not the liberator that the Jews expected, because, Scripture says, it was necessary for the Messiah to suffer, die, and rise again on the third day before reigning. Although, Jesus taught the forgiveness of sins and the spiritual life, obeying by grace and understanding, the most preeminent elements of the Law of Moses. In contrast to this, Bacchus is said to have been a kind of liberator; where said liberation entailed and was summed up in an ecstasy of the normal being, through madness, debauchery and wine.

On the other hand, it is known, thanks to Matthew Levi -one of the 12 apostles-, that Jesus spent his childhood in Egypt, which, the detractors of the story of Jesus say that it was taken from the story of Bacchus (Dionysus), since, according to Greek mythology, he was sent to Africa in his childhood. However, the story does not say that he was sent to Egypt but to Ethiopia. Jesus is said to have passed through India at some point before beginning his ministry in Galilee, though his actual childhood is unknown. Critics affirm that this is the same as the story of Bacchus, since, officially, his childhood is not known, but they affirm that he was in India. As far as symbols are concerned, Jesus has many, including the cross, and later the fish -adopted by Christianity, not by Jesus-, but especially bread and wine. For its part, the symbols of Bacchus are antagonistic, since it is identified with the serpent and the bull, precisely the anti-Jewish emblems, especially since the bull usurps the honor of

Jehovah (Abraham took the symbol of the bull, called Aleph, and dedicated it to to Jehovah, around 2000 BC).

For this reason, if Bacchus has the bull, the serpent, the ivy and the wine as slogans, and sometimes the fig tree, it is a bad example to relate him to Jesus. Jesus is not the wine, but he used this symbolism, which resembles blood, to seal the New Covenant, as Abraham had done with Melchizedek 20 centuries before, since it was his way of sealing a pact - along with the bread -. Now, Jesus encourages chastity and a pious life, and he himself lived in chastity. Instead, Bacchus incited drunkenness and lust, and lived in orgies, in addition to having been married to Ariadne, after she had been abandoned by Theseus. In that order of things, Jesus did not identify himself with the fig tree either, but the fig tree is the symbol of the nation of Israel, which emerged as such even before Moses, around 1900 BC.

Regarding the theme of wine, Jesus is presented as " *the true vine* ", " *the resurrection* ", " *the life* ", " *the light that has come into the world* ", while Bacchus was known as the god of wine and with the symbol of the vine of the vineyard (the false vine). What does this mean? If Jesus spoke of the " *true vine* ", it was because he assumed that there was a vine that was not true: "*I am the true vine, and my Father is the husbandman. Every branch in me that does not bear fruit, he takes away; and all that bears fruit, he cleanses it, so that it bears more fruit. You are already clean because of the word that I have spoken to you. Stay in me, and I in you. As the branch cannot bear fruit by itself, if it does not abide in the vine, so neither can you, if you do not abide in me. I am the vine, you are the branches; he who abides in me, and I in him, this one bears much fruit; because apart from me you can do nothing.*" (Apostle John 15:1-5) The apostle John records this dialogue, and it is understood that Jesus speaks of giving a result based on the message that he came to teach, expecting from his disciples that they would bear "more fruit", that is *to* say, more results than he himself gave in his 3 and a half years of work.

It is obvious, Jesus used as references the things that the people of the surroundings knew, especially the tillage of the Earth. He also knew of the Greek influence, because Galilee was a point of convergence where Greece played a large part in the culture – that is why they called the region « *Galilee of the Gentiles* ». Regarding chastity and purity, Jesus does not resemble any of the characters attributed to him, as an invention that appeared under the influence of all of them. He is masculine and incites the obedience of the commandments of Moses, while Bacchus is represented as effeminate or half man and half woman, in the same way that occurs with Attis. Although, Jesus surprised by his wisdom and eloquence, having been tested in everything, but without having sinned in anything. Instead, Bacchus was driven mad by Hera and wandered for a while, while Demeter was cast out of the Olympian pantheon.

Jesus taught to reject wealth and materialism, while Dionysus could empower others to have the power to turn everything he touched into gold. This is one of many contradictions: « *Take heed, and keep yourselves from all covetousness; because the life of man does not consist in the abundance of the goods that he possesses.*» (Doctor Luke 12:15) And with many more words Jesus taught that " *It is easier for a camel to go through the eye of a needle, than for a rich man to enter the kingdom of God.*" (Apostle Matthew Levi 19:24) Another matter of debate in this regard is the fact that Jesus rose from the dead three days after he died, when he was about 33 years old. As a parallel, some see Dionysus as being resurrected in his childhood after being mutilated by the Titans on Hera's orders, however, that apparent resurrection did not happen at all as it is told. It is not that I say that it was true, but that the myth is distorted, since the titans ate almost the entire child, except the heart, which Zeus put in the womb of Semele.

That, therefore, is not a resurrection in the manner taught in Judaism. In fact, Jesus was not the first person to be resurrected nor the only one in Israel. Before him many people were resurrected in Jerusalem (Apostle Matthew Levi 27:52-53), he himself brought many people back to life -among them his friend Lazarus-, and his disciples also continued to do so, because this was one of the keys to the power and mission of Jesus Christ: « *Heal the sick, cleanse lepers, raise the dead, cast out demons; freely you received, freely give."* (Apostle Matthew Levi 10:8) Moreover, during the time of the reign of Israel, centuries before Jesus, the prophet Elisha raised a young man from the dead: "When Elisha came to the house, behold, the child was lying dead *on his bed. He then entered, closed the door behind them both, and prayed to Jehovah. Then she went up and lay down on the child, putting her mouth on his mouth, and her eyes on his eyes, and her hands on his hands; so she stretched out on him, and the child's body became warm. Turning then, he paced up and down the house, and then climbed up, and stretched out on him again, and the child sneezed seven times, and opened his eyes. "* (2nd Kings 4:32-35, Old Testament)

It is obvious that Jesus identified himself with the Resurrection and with Life, since that was his mission: « *The thief only comes to steal and kill and destroy; I have come that they may have life, and that they may have it to the full."* (Apostle John 10:10) What is that abundant life? That Jesus has come to eradicate death, bring the Resurrection upon all men and grant us an Eternal life, without death, without aging, without disease and without deterioration, but in due time, "until everything is fulfilled." (Apostle Matthew Levi 5:18) Continuing with these apparent parallels we find that Jesus proclaims himself as the lion of the tribe of Judah. On one occasion, Dionysus became a lion, in a boat, but it is not his characteristic. This is not a similarity, moreover, Jesus is called a lion because the lion is a foreign animal – in Israel -, which is the maximum in sovereignty. It is the best example to characterize him, and because Judah, the tribe with which he is characterized, understood that the lion was his emblem, that is, Jesus said that he was the representative and king of Judah, the Israelite caste from which the monarchy comes. of the offspring of Jacob.

You don't have to go any further to see other disparate examples between Jesus and Dionysus, such as that Jesus is the "*Prince of Peace*", while Dionysus, according to Greek literature, had a certain destructive nature. Now, regarding his nature, Jesus is the firstborn of all Creation and lord of the entire universe, while Dionysus was merely one more deity of the Olympian pantheon. Jesus taught that his blood is the new alliance and his flesh is the body or organization that precedes it, recalling the alliance as such, which was already made in the days of Abraham and Melchizedek. As I have already said, he used bread and wine as symbols, basically why it is said that he was the creator of wine, although it is never mentioned that he turned water into wine. Jesus Christ is the Messiah of Israel, his own nation, while Bacchus was a foreign god. Although, the only logical parallels are that Jesus had a stepfather, but died when he was 14 years old, while Bacchus lived with his adoptive father.

ATTIS. This Greek figure, of course, is another apparent emblem of which it is said, ignorantly, that the Christians invented Jesus – since Jesus, as we have already seen, is more than a historical character. To begin with, the first literary reference to Attis is the subject of one of Catullus's most famous poems, but it seems that the cult of Attis in Rome did not couple with the pre-existing cult of Cybele until the early Empire, that is, it came to be something rather recent. Initially Attis was a local Phrygian demigod, who became more specifically known on the Agdistis mountain, who was personified as a daemon (demon). Identifications of the name Atys go back to the 19th century and are found in Herodotus, but as the historical name of the son of Croesus, in "*Atys the sun god, wounded by the tusk of the boar*", and as a deity of life, death and resurrection as described by James Frazer, are wrong.

We already know more about the nature of Jesus, so it makes no sense to claim a source for the myth of Attis, who was the lineage of the daemon Agdistis, who initially had both male and female attributes, and who was said to have had his organ cut off. masculine and they threw it away, growing on the site where an almond tree fell, from which Attis later emerged. As in the case of other Greek characters such as Dionysus, Hephaistos or Zeus, he was raised outside the paternal and maternal womb, being cared for by a ram – an emblem that does not correspond to Jesus Christ at all, but rather is his opposite. There is also talk of Attis's adoptive parents, who sent him to Pesino, where he was to marry the king's daughter. Where is the parallel with Jesus? Of the particular things of Attis, it is said that he went mad and cut off his genitals, which, apparently, led to his death. Attis was reborn as an evergreen pine tree. This alleged rebirth was celebrated on March 25 and, although, it is more similar to Dionysus than to Jesus.

Accounts of alleged "resurrections" are also seen in the story of Dumuzi and the tree of Nimrod. It is not said of Jesus that he was necessarily dead for three days, but that he died, *"was made alive"*, and in that condition he went down to Hades with the purpose of taking ownership of the keys of death and giving testimony to the dead and to the fallen angels. Other than this, Attis bears no resemblance to Jesus or vice versa. These aforementioned characters were reborn from things already created and transformed by the apparent work of other gods, who in many cases could not help them return to their normal state: «For Christ also suffered *once for sins, the just for the unjust , to lead us to God, being indeed dead in the flesh, but made alive in spirit; in which he also went and preached to the imprisoned spirits, those who once disobeyed, when once the patience of God waited in the days of Noah, while the ark was being prepared, in which few people, that is, eight, they were saved by water.* » (First Letter of Peter 3:18-20)

KRISHNA. It is said in Hinduism that Krishna was one of the many avatars of the god Vishnu, and in Krisnaism, Krisná -as it is said among them- is the main form of God, from whom Vishnu and the other gods emanate. What is known as God, or Brahma, is a Hindu concept, which was not known in Israel, and which took time to be known in Rome at that time. According to the British Sanskritologist Monier-Williams (1819-1899), the Sanskrit term krishna means: black, dark, a type of demon or spirit of darkness, behaving in a dark way, name of a hell, the dark age. According to the jainas, Krishná is one of the nine black vasu devas. According to the Buddhists, he is the chief of the black demons, who are enemies of the Buddha and the black demons. Others say that it is the name of an asura (demon), etc.

All resemblances with dark and diabolical beings does not seem to be an example of Jesus Christ, who, precisely, cast out demons, and who clearly said to the Jewish religious leaders who said that he was a minister of Beelzebub: "I do not have a demon, rather I honor *my Father; and you dishonor me*." (Apostle John 8:49) Basically the term Krisna is associated with the color black, which is why he is called The Dark Lord, a quality and designation opposite to Jesus. Krishna was said to be a shepherd god. Although, Jesus never said that he was a god –although it was understood by his way of speaking about the Father-, nor a shepherd or shepherd god. Moreover, the references to the polytheistic gods were totally contrary to Hebrew theism, and in that their abysmal difference was characterized in all areas. For the Israelites there is only one God, Jehovah, and the rest are usurpers and fools.

Precisely one of the arguments of the Jews to deny Jesus is the fact that Christians consider them "the Son of God", and that is clearly anti-Jewish, since there is no reference to Jehovah having a son. Furthermore, the Torah says that God is the only God, and there is no other but Him. Those who called Jesus "Son of God" or "a god" were non-Jews. His disciples, at the time, did not understand the significance of this, precisely because they had no knowledge that God had a son, since they expected it to be a liberating king from the tribe of Judah, who must be the Anointed of God. to govern all Israel and free it from the oppression of foreign peoples. When some spoke of Jesus they came to identify him as "a son of God", but not as "the son of God", since by that rule we are all children of God.

Now, it was argued from the study of the Tanak that it was very probable that the Messiah had a divine and immortal nature, but this was not understood in its context – in fact, it took the disciples a long time to understand and assimilate it. Although, although both are called shepherds, Krisná was shepherd of cows, while Jesus is shepherd of men, whom he affectionately calls sheep. In addition, Jesus, knowing all these traditions because he was characterized by his great erudition, says: « *I am the good shepherd; the good shepherd lays down his life for the sheep. But the hireling, and who is not the shepherd, to whom the sheep are not his own, sees the wolf coming and leaves the sheep and flees, and the wolf snatches the sheep and scatters them. So the hireling flees, because he is a hireling, and he does not care about the sheep. I am the good shepherd; and I know my sheep, and mine know me, just as the Father knows me, and I know the Father; and I lay down my life for the sheep.*" (Apostle John 10:11-15) If he speaks of being " *the good shepherd* ", it is because he exposes that there are others who are not good shepherds or who are impostors. No god has ever given life for humanity, not even some of these supposed epithets, which were before Jesus.

In addition to the fact that the concept of reincarnation is contrary to the Hebrew traditions, it is indicative of satanic actions, since human possession is considered a violation of natural laws unless it is of your own and only body: «And in the way *that It is established for men that they die only once, and after this the judgment, so also Christ was offered only once to bear the sins of many; and he will appear a second time, unrelated to sin, to save those who are waiting for him."* (Letter to the Hebrews 9:27-28) Well, Krisná belonged to the tribe of the iadus, of the dynasty of the Moon, which clashes with Judaism, since the moon was a naturally negative symbol. It is said that Kriná was warned by a sage named Nárada Muni that he would die at the hands of a son of his sister Princess Devakī with her husband Vasudeva. Apparently there is a certain resemblance, but Jesus was not killed by a relative, but by a foreign people.

Although, the Krisna culture does not fit with this parallelism, but a root can be found in Hinduism, where it is said that Vishnu incarnated in Krisná. However, although it sounds like a Jesus-like account, in one small respect the relationship of the Holy Spirit and Jehovah with Vishnu is totally out of context. Vishnu has been important to Hindus, but he is not primal or superior to Brahama. In the case of Jesus, he made it clear that there is no sovereign over the Heavenly Father, from whom he came. Jesus was not an incarnation of Jehovah or of the Heavenly Father or of the Holy Spirit, but rather a representative: « *No one has seen God at any time; the only begotten Son, who is in the bosom of the Father, he has made him known.*" (Apostle John 1:18) Jesus came to introduce us to the Father, a concept only smacked of in Scandinavia, where they called Odin the All-Father. However, although there were similar epithets, their characteristics, life and history were totally different.

What it was about, in the mission of Jesus when teaching about the Heavenly Father archetype, was for humanity to know that there is paternity with a God, who is the Creator of everything that exists, while in polytheism the gods were of a higher caste and used mortals as they pleased: «*The God who made the world and all things in it, being Lord of heaven and earth, does not dwell in temples made by human hands, nor is he honored by hands of men, as if he needed something; for he it is who gives to all life and breath and all things. And from one blood he has made the entire lineage of men, so that they may dwell on the whole face of the Earth; and he has fixed for them the order of the times, and the limits of their habitation; so that they seek God, if in some way, by feeling, they can find him, although he is certainly not far from each one of us. For in him we live, and move, and have our being; as some of your own poets have also said: Because we are his lineage. Being, then, a lineage of God, we should not think that Divinity is similar to gold, or silver, or stone, sculpture of art and the imagination of men.»* (Acts of the Apostles 17:24-29)

The other characteristics of Krisná only make things worse for those who are looking for an argument that can affirm that Jesus is a mythological character of Hindu origin: he misbehaved as a child, had affairs with shepherdesses, was a tricky hero, killed giants who wanted to assassinate him in his childhood, raised his kingdom in his time, moved notoriously in a polytheistic environment, had 16,108 wives, had thousands of children, was accidentally killed by a hunter, died at the age of 125. Jesus has none of these characteristics: they are all the opposite of him. Another aspect to note is the Vedic reference that Krisná is an eternal being, without birth or death, who would have adopted a temporary body to be able to be born and die on Earth, but simultaneously he would be eternally present on his spiritual planet. His devotees consider that surrendering to him (having krisná chaitania, "Krishná consciousness"), leads them to spiritual perfection and eternal happiness.

Obviously, some would have postulated that this is similar to the Christian position of Salvation and Eternal Life. However, Jesus, although he speaks of spiritual progress in everyone, until reaching his "*fullness*", emphasizes Hope in the promise of the Resurrection of all the dead to acquire an "*incorruptible*" and "*immortal*" body. Christ also affirms that he will return with his Father's Kingdom to rule on Earth, and that he is now in his celestial city, which comes from some distant part of the universe. Although Krisná died and was seen as an etheric being, this does not happen with Jesus, who is physically alive –forgive the redundancy-, as he showed his disciples after resurrecting: «Look at my hands *and my feet, that I myself am; handle, and see; for a spirit does not have flesh and bones, as you see I have.*" (Doctor Luke 24:39)

MITHRA. Known as the god of sunlight, Mithras appears as another of the supposed beings that inspired Christianity to give rise to the figure of Jesus. As you can already see, all these characters have nothing to do with Jesus, who is clearly superior to all of them, and clearly a historical figure that has marked humanity. Well, Mithras, a Persian and Indian personality who passed into ancient Rome, is depicted as a young man, in a Phrygian cap, slaying a bull with his bare hands. During the Roman Empire, the cult of Mithras developed as a mystery religion, and was organized in secret societies, exclusively for men, of an esoteric and initiation nature. It enjoyed special popularity in military environments. He forced honesty, purity and courage among his followers. From archaeological findings it is known that the cult of this being is a religion of Persian origin, adopted by the Romans in the year 62 BC. C., which competed with Christianity until the fourth century.

There are really few texts written by Mithraist authors. Some paintings and inscriptions have survived, as well as descriptions of this religion by its opponents, including Neoplatonists and Christians. Much of what has circulated about this Mithraism has been based on the theories of a Belgian scholar named Franz Cumont. His work entitled "The Mysteries of Mithras", published in 1903, led to assertions by the School of the History of Religions, in the sense that Mithraism had influenced some practices of the incipient Christianity, but this being merely his point of view. view. Over time, this led to more popular academic circles - and along with the seed that Kersey Graves had planted with his book (considered pseudo-history by Christian and non-Christian scholars alike) "The World's Sixteen Crucified Saviours", in 1875 - a very elaborate urban legend was formed about an alleged virgin birth of Mithras, as well as an alleged death and resurrection of this character, and several other points that closely relate his life to that of Jesus of Nazareth.

This hypothesis has no historical foundation, and it does not correspond to the data available on Mithraism, much less with the history that can be extracted from the historical findings on Mithras. The own ideas of the revolutionary Alfred Loisy have no support against the basis of Hebrew culture. Loisy was one of the precursors who said that Christianity was based on a pagan Judaism that absorbed the other cultures of Europe, but this statement is no less than ridiculous, if we consider that Judaism is precisely anti-pagan. Even with that, the hypothesis remained strongly rooted until today. For example, the secret of Mithraism was not the faith but the rites, something acceptable to Israel, if one is unaware of Jewish traditions, but not applicable to Christianity. The brotherhoods of Mithras admitted only men and not women who did not participate in the functions of the cult, contrary to Christianity.

Moreover, the paganism of the trinity of Athanasius that Catholicism imposed does have a similarity here, since the Father is Zeus-Ormazd, the son is Mithras and the other is the bull. However, in historical light, Zeus is associated with Jehovah's enemy Baal the Canaanite, and the bull is also a usurper of Jehovah. The trinity itself is anti-biblical and discarded in Judaism, as it was with the early Christians - later adopted by Constantine. About other references, such as the myth of the bull sacrifice, almost everything is ambiguous. Said symbolic sacrifice during the rite at the hands of Mithras had as its purpose the redemption and immortality of the followers. On the sacrifice of the bull (representing Mitra) rested the balance of the world and the salvation of men. Although, this seems more like a plagiarism from Mithraism to original Christianity than the other way around. The rest of the association with the bull is complete "heresy" - if we use the valuation that Rome gave to this Latin word.

Knowing the rejection of the law of Moses to the worship of the stars, in the rites of Mithras the 7 stars are worshiped, which clearly are similarities of the " *principalities and powers* " that Paul pointed out as " *hosts of wickedness in the places celestial* .» (Paul's Letter to the Ephesians 6:12) Then there is the fact that the Israelites consecrated the seventh day, without relation to the stars –because it was forbidden-, while those of Mithras, as in Catholicism, worship the sun. Also, as in Rome, those of Mithras had leaders as fathers, which in Judaism and early Christianity were allegorically intended for Israel's leaders such as Abraham, Isaac, and Jacob, although Jesus certainly did not accept dealings with anyone at all. way of Father: « *And do not call your father your father to anyone on Earth; for one is your Father, who is in heaven.*" (Apostle Matthew Levi 23:9) Supposedly Mithras baptized his believers and promised atonement for sins by the effect of bathing. Only in this cult was the imposition of a sign on the forehead attached to baptism, as in the Catholic Church.

In any case, baptism was not something new in the Jordan in the days of John the Baptist. There are references to baptism even in Scandinavia, either for initiation, symbolic birth or for liberation, because that is precisely the basis of its meaning: "immerse" (Greek: "baptos"). Other authors also see a parallel between Christianity and Mithraism in that the birth of Mithras was celebrated on December 25, but Jesus was not born in winter. Rome, always the culprit in all this string of misinformation, adopted December 25 for the birth of Jesus, to fit in with its paganism and polytheism. That is why the attributes of the pater -the highest level of initiation in Mithraism- were the Phrygian cap, the rod and the ring, very similar to the miter, the staff and the ring of the Catholic bishops. Catholic Rome is a disguise for Egyptian, Babylonian and Greek worship.

Since the Council of Nicaea in 325 AD, when Catholicism was born, the true message, teachings and principle of Jesus and his apostles had already been lost. Precisely, already speaking of the birth of Jesus, "*There were shepherds in the same region, who kept watch and kept watch over their flock by night.*" (Doctor Luke 2:8) Shepherds in the middle of a winter night? This happens due to the lack of culture, because no one grazes on winter nights because they are icy. In addition to being a deity absorbed by the Romans, Mithras is a Vedic figure from India, a supposed god. According to the Bhagavata Purana, he is the god who controls the intestinal movement (in Sanskrit the term mitra means "friend"). To understand the importance of proselytizing polytheism, one must focus on the mythological roots of all these personalities. Mitra is, according to another version, one of the Aditya, the children of the goddess Aditi. According to some sources, his brothers may be seven or eight, although other references go so far as to say that up to thirty-one.

Aditya indicates their classification as solar and/or sky gods. According to the Rig Veda, Aditi is a female deity, mother of all gods, wife of Kashyapa, and daughter of Daksha, a minor progenitor god of the universe. It is said that she contains everything, and could be considered as "nature" or "primal creator goddess", which is the same as the common concept of "mother Earth", which is popular in naturalism. It is important to note that to believe in Jehovah or accept monotheism, society tends to shut down, but to believe in other divine figures such as Mother Nature or in Catholic Saints or Virgins of stone, wood, porcelain or plaster, everyone is fervent devotees. Mithras is a secondary god of the sun, and this, like the rest of his characteristics, has nothing to do with Jesus. From the outset, Jesus never identified himself with the sun, besides being, this idea, a violation of the commandments given to Moses, and which the prophet Jeremiah reproached Israel.

Because they scolded the people, they said to Jeremiah: « *The word that you have spoken to us in the name of Jehovah, we will not hear from you; but we will certainly put into practice every word that has come out of our mouths, to offer incense to the queen of heaven, pouring libations on her, as we and our fathers, our kings and our princes, have done in the cities of Judah and in the marketplaces. from Jerusalem, and we had plenty of bread, and we were glad, and we saw no evil. But since we stopped offering incense to the queen of heaven and pouring out libations on her, we lack everything, and we are consumed by the sword and by famine. And when we offered incense to the queen of heaven, and poured libations on her, did we make her cakes to worship her, and pour libations on her, without the consent of our husbands?* » (Prophet Jeremiah 44:16-19) Because of this stubbornness and disobedience the Israelites were later delivered into the hands of their enemies.

These words and these events were later recalled by the apostle: « *And God withdrew, and gave them over to worship the host of heaven; as it is written in the book of the prophets: Did you offer me victims and sacrifices in the desert for forty years, house of Israel? Rather, you carried the tabernacle of Moloch, and the star of your god Renfan, figures that you made for yourselves to worship. So I will carry you beyond Babylon.*" (Acts of the Apostles 7:42-43) If Jehovah punished the Israelites for these practices, and precisely for worshiping these gods that they compare to Jesus, how were the Jews going to create a Messiah who represented everything that violated the laws? Jehovah's codes, if Jesus himself commanded the people to obey God? That is why he affirmed: «*So whoever breaks one of these very small commandments, and thus teaches men, very little will be called in the kingdom of heaven; but whoever does and teaches them, he will be called great in the kingdom of heaven.*" (Apostle Matthew Levi 5:19)

And on another occasion he said to a rich young man who was asking him how to inherit the Kingdom of God: «*But if you want to enter life, keep the commandments. He said to him: Which ones? And Jesus said: Thou shalt not kill. You shall not adulterate. You shall not steal You will not give false testimony. Honor your father and your mother; and, You shall love your neighbor as yourself.*" (Apostle Matthew Levi 19:17-19) Subsequently, in a summarized way, Jesus exposed his approval of Jehovah's law: «*And one of them, an interpreter of the law, asked to test him, saying: Master, what is the great commandment? In the law? Jesus said to him: You shall love the Lord your God with all your heart, and with all your soul, and with all your mind. This is the first and great commandment. And the second is similar: You shall love your neighbor as yourself. On these two commandments hang all the law and the prophets.*" (Apostle Matthew Levi 22:35-40)

Other of the parallels that try to associate Jesus with Mithras, is the fact that he is related to oaths, promises, contracts, honesty, friendship and meetings, as well as considered as the soft sun of the Sunrise. However, what does that have to do with Jesus? The only thing here that Christ does teach is the love of friendship and fellowship, in addition to obviously teaching, through his example, honesty, but to start from these examples it is not necessary to copy anyone. Jesus is not the sun, much less the morning sun, although he has power over the morning star: "*To him who overcomes and keeps my works to the end, I will give him authority over the nations, and he will rule them with a rod of iron, and they will be broken like a potter's vessel; as I also have received it from my Father; and I will give him the morning star.*» (Revelation of John 2:26-28)

Jesus is obviously not a star as such, and furthermore, if he gives this star to the victor, does he give himself? Now, it is obvious that the concept of the morning star cannot be something literal, since it itself says: «*I Jesus have sent my angel to testify to you these things in the churches. I am the root and offspring of David, the bright morning star."* (Revelation of John 22:16) If Jesus associates being descended from David with the symbol of the morning star, he may mean the star of David, again a symbol. In the Hebrew culture angels are represented as stars. In fact, it is said that Satan was the morning star, something merely referring to a title that he held and which Jesus takes away from him. The mere concept refers to being the origin of things, the light that first took place, since Jesus is known as *"the image of the invisible God, the firstborn of all creation."* (Paul's Letter to the Colossians 1:15) We are talking about Jesus having the greatest titles that any deity has ever had and having an origin and nature even greater than all of them.

What does all this mean? that the gods that supposedly were before him, are considered by Jehovah as deserting angels, whom Jesus ridiculed with his arrival, ministry, death and Resurrection: «And you, being dead in sins and in the uncircumcision of your *flesh He gave you life together with him, forgiving you all your sins,* <u>*annulling the handwriting of decrees that was against us*</u>*, which was contrary to us, taking it out of the way and nailing it to the cross, and* <u>**stripping principalities and powers, he exhibited them publicly, triumphing over them on the cross .**</u> » (Paul's Letter to the Colossians 2:13-15) We can continue to find inconsistencies about the apparent parallels of Jesus with these deities, such as the reference to the Rig Veda, where the role of moon god or Chandra, later assigned to Shiva, is mentioned, which received Mithras. The moon, as I have said before, is an anti-Hebrew symbol, as a cult – although they were based on the lunar calendar. As we will see with the next candidate, the story of Jehovah's enemy gods was not a myth or a figure of speech, since it dealt with characters who, as the astronaut theorists of the past affirm, descended from heaven to Earth in very ancient times. ancient and became gods.

ZOROASTER. Now, out of these divinities, the only human personality that is attributed to be a predecessor of Jesus is Zoroaster. It is said that he was born of a virgin, that he was baptized in a river, that he began to preach at the age of 30, that he was tempted in the desert by demons, that he restored sight to a blind man, that he revealed the mysteries of heaven, the Resurrection of the soul and Salvation, and that he was celebrated by eating his body as a symbol – a kind of Eucharist as developed by Catholics. And well, although he is a human, in fact, a prophet, as Jesus is also, none of the above can be supported historically. The reason is simple, little or nothing is known about him directly, and the few references that are known are shrouded in mystery and legend.

As official references to Zoroaster, none of the above claims are recorded, except that he was one of the earliest precursors of monotheism and strongly influenced ancient Persia and Afghanistan. He spoke of practically the same things that were later glimpsed also in Christianity and Islam, which were believed among the Hebrews, and which were similar to ideas that Pharaoh Akhenaten came to believe, and the Romans at a time when they considered that indeed there was only one God, called Numen. So, it is more likely that all these supposed parallels are an invention of those who try to discredit Jesus of Nazareth. The teachings of Zoroaster or Zarathustra, depending on the language used, are compiled in what remains of the Avesta, where he exposes his beliefs in the form of songs.

HORUS. His name means "the Exalted", a very proper title for a deity. Jesus means "Salvation", because he came humbly to save men from their sins. Horus is directly associated with the Greek Apollo, whose nickname was Phoebus, the sun god, successor to Helios. Horus was identified as a falcon or falcon-headed man. Since the Old Kingdom, the pharaoh is the manifestation of Horus on Earth, although when he dies he will become an Osiris, and will be part of the so-called creator god Ra. During the New Kingdom it was associated with the god Ra, as Ra-Horajty. It is a core part of the Great Ennead (the basic and primeval pantheon of Ancient Egypt). It is part of the Osiriac triad: Osiris, Isis, Horus. For these reasons, Catholicism cemented the idea of a Triune God: God the Father, God the Son and God the Holy Spirit, an idea categorically rejected by Judaism, and which cannot, in any instance, be supported by the Bible, since they are polytheistic ideas.

Certainly the Israelites absorbed a lot of Egyptian influence during their stay in that nation, but for that very reason they had to spend 40 years in Sinai debugging those ideas and absorbing the ones that Yahveh (Jehovah) brought them. However, his name, Israel, was clearly Hebrew: Ish-Ra-El (Man-Sees-Most High), by God's promise to Jacob, Abraham's grandson, to whom he changed his name: «And God said to *him: Your name is Jacob; Your name will no longer be called Jacob, but Israel will be your name; and he called his name Israel.* » (Genesis 35:10. The Torah) Centuries after this, his descendants entered Egypt to live. The word "Ra" is as old as it is enigmatic, but the fact that Horus is often associated with one eye, or Ra, is because he sees everything. Hence the phrase "he who sees everything" comes from. The word "El" also has its origin in Canaan, since that voice came from Mesopotamia, specifically from Acad, where the gods were called "Ilu".

From that well-known word, to describe deities, but Allah or El, because they were clearly the ways of calling a divinity, for something else. Subsequently, Jehovah took the trouble to change the traditions and the value that words had in all these regions. Some err again trying to find in Jesus a Hindu or Egyptian -or even Greek- origin, especially trying to associate him with a complete opposite of himself: a god of war. Shortly after he was born, Horus, son of Osiris, was hidden by his mother Isis and left in the care of Tot, god of wisdom, who instructed him and raised him until he became an exceptional warrior. Clearly, Moses was also hidden in his infancy, and anyone who has read the Book of Jasher will know that Abraham was also hidden in his infancy so as not to die at the hands of Nimrod.

What's more, less is known about the childhood of John the Baptist, or his predecessor, the prophet Elijah, than about Jesus himself. Jesus was never trained to be a warrior or to fight or enter into combat, indeed, King David said about him: "The Lord said *to my Lord: Sit at my right hand, until I make your enemies your footstool.*" (Psalm 110:1. Written by King David, or by Asaph) Jehovah was telling Jesus to take command while Jehovah eliminates his enemies, so that then Jesus can take over the Kingdom. There is also evidence that the father of Horus, Osiris, was assassinated and after his failed resurrection he became a judge of the underworld. Does this have anything to do with the story of Jesus or Heavenly Father? In any case, Horus is a warrior god by nature and fully fledged, a solar god and a leader of all Egypt – although he began only with Lower Egypt, until he defeated his uncle Seth and banished him.

In addition to these ideas, the representations of Horus are varied and depending on the area of Egypt and the mandate in which each area was developed. Almost every epithet related to battle and the passing of the sun is applied to Horus, even in his own childhood. Clearly there is virtually no relationship between this deity and Jesus of Nazareth. Horus was already performing his functions as a child, but Jesus only showed his precocity and wisdom, since " *his hour had not come* ", as his disciple John affirmed. It was after his baptism in the Jordan that Jesus began his ministry, and he did so in what was left of Israel during the reign of Tiberius in Rome. It is also worth emphasizing that it was absurd that a tradition of Jesus was born in Judea in the first century in the way that its opponents try to expose. Another very different thing are the inventions that Rome generated in the fourth century, when giving birth to the universal religion (Catholicism), where it took ideas from all the peoples of yesteryear and created a mixture that obscured the true message of Jesus and his disciples. in all those regions.

The problem of not knowing ancient history, Hebrew literature, archeology and mythology is that one can reach these gross inconsistencies and fallacies, very delicate, since they call into question the identity of the most important character in all of our history: Jesus, the Messiah. Delving into the past we must recognize that there were angels who deserted from heaven and came down to Earth to live in their own way. They, for this action, were removed from the Kingdom of Heaven. Those angels, which today, due to our advanced vocabulary, we call extraterrestrials, became gods, mainly in Egypt, Babylon and India. Later they settled in Canaan and then in Greece, until the time when Jehovah began direct war against them. Since then, it is possible that they emigrated to Scandinavia, South Africa, the Far East and America, fleeing the battles that Jehovah and his armies waged against them.

Because of it, is imperative that one be aware of all these things and investigate them diligently. Jesus came, when these gods had already been practically overshadowed, to teach humanity that death is not the end and that a resurrection from the dead and an Eternal Life awaits us, when a judgment on all men takes place, for determine the fate of each. Although, Jesus could not come before because of the sovereignty of those gods, which had their greatest apogee or Golden Age, from the days of the Garden of Eden, until the fall of the Egyptian Eighteenth Dynasty, approx. This is how Paul explained it, referring to this period of time without God, as "death": "*Nevertheless, **death reigned from Adam to Moses**, even over those who did not sin in the manner of Adam's transgression, which is a figure from the one to come.*" (Paul's Letter to the Romans 5:14)

HISTORICAL JESUS

As a touch of grace in this religious war, the place of Jesus Christ has come to be secondary, when it comes to giving him prominence, but primary when it comes to wanting to unmask Catholicism. As I said before, Rome's problem was to take the history of the Hebrews and use it to its advantage, since the Jews were no longer on the scene to complain about this arbitrariness. Why weren't they? Because they had been expelled from Judea by the Romans, after the three Judeo-Roman wars of the first century, and they were also expelled from Rome, as Luke the physician wrote: «After *these things, Paul left Athens and went to Corinth. And he found a Jew named Aquila, a native of Pontus, recently come from Italy with his wife Priscilla, because <u>Claudius had ordered all the Jews to leave Rome</u>*." (Acts of the Apostles 18:1-2)

This is confirmed by Gayo Suetonio Tranquilo, when making Claudio's biography, around the year 121 d. C.: «*he expelled the Jews from Rome*», adding that the reason was because «*they continually caused riots*», for being harassing the Christians. This appears in his work called "On the Life of the Caesars." It is also known that the Jews lost the wars against the Romans in ancient Judea, thus beginning the deterioration of their control over the Holy Scriptures, which remained in Babylon with relevant texts such as the Septuagint –although the Sanhedrin wanted to avoid this Greek translation to prevent the Jews from knowing that Jesus is the Messiah that the prophets of Israel had warned would come.

The public history of Jesus was greatly clouded from the beginning of his ministry in Judea, Galilee and Caesarea. His main opponents were the blackest members of the Sanhedrin, who called him Beelzebub and the Samaritan magician. Subsequently, with the disappearance of Jesus, the work of his missionaries and those who were sent was hard, precisely against the Jews everywhere where his message reached, since half accepted that Jesus was the Envoy, but others denied it and made a strong War on Messianic Jews. After Jerusalem fell in 135 AD, the Jews disappeared from the map –literally- and were forced to emigrate in all directions and remain in exile. Jerome took the literature of the Hebrews and translated what is now called the Bible into Latin, but without the permission or approval of the Jews, to whom this story belonged.

It was Constantine the Great who legalized the Christian religion by the Edict of Milan in 313 AD, giving birth in his days to Catholicism, which was powerfully organized when he convened the First Council of Nicaea in 325 AD, where Christianity was granted legal legitimacy in the Roman Empire for the first time, although in itself it had nothing to do with primitive Christianity, known as Judeo-Messianism. Around the year 112 AD, Pliny the Younger, imperial legate in the provinces of Bithynia and Pontus (located in present-day Turkey), wrote a letter to the Emperor Trajan to ask him what he should do with the Christians (Pliny the Younger. Epistle X, XCVI, C Plinius Traiano Imperatori), many of whom he had had executed. In that letter he mentions Christ three times about Christians. On the third occasion he says that the Christians " *claimed that all their guilt and error consisted in meeting on a fixed day before dawn and singing to alternative choirs a hymn to Christ as a god.*"

Soon after, around the year 116 AD, the Roman historian Tacitus wrote his "Annals" and also mentioned Jesus. In Book XV of the Annals, Tacitus narrates the terrifying fire of Rome in the year 64. It was suspected that the fire had been ordered by the Emperor Nero. Tacitus writes that "*in order to put an end to the rumors, Nero presented as guilty and subjected to the most elaborate torments those whom the vulgar called Christians, hated for their ignominies. The one from whom they took their name, Christ, had been executed in the reign of Tiberius by the procurator Pontius Pilate; the execrable superstition, momentarily repressed, broke out again not only through Judea, the origin of evil, but also through the City ...»* (Annals, 15:44:2-3) Later, in the second half of the second century, the writer Luciano de Samosata, a native of Syria, referred to Jesus in two burlesque satires ("On the death of Pilgrim" and "Proteus").

In the first of them he speaks of Christians as follows: « *Afterwards, by the way, of that man whom they continue to adore, who was crucified in Palestine for having introduced this new religion into the lives of men... Besides, their first legislator told them convinced that they were all brothers and so, as soon as they commit this crime, they deny the Greek gods and instead adore that crucified sophist and live according to his precepts.* » At the end of the first century, the Syrian Mara ben Sarapion referred to Jesus in this way in a letter to his son: «*What benefit did the Athenians obtain by killing Socrates, a crime that they had to pay for with famines and plagues? Or the inhabitants of Samos when burning Pythagoras, if their country was soon drowned in sand? Or the Hebrews when executing their wise king, if soon they saw themselves deprived of their kingdom? A god of justice avenged those three wise men. The Athenians starved to death; those of Samos were swallowed up by the sea; the Hebrews were killed or expelled from their land to live scattered everywhere. Socrates did not die, thanks to Plato; neither did Pythagoras, because of the statue of Era; nor the wise king, thanks to the new laws promulgated by him.*»

Although, the fact that there was no Messiah in the first century, except Jesus, is irrefutable. It is not that they existed in some other era, but that important personalities did exist, many as Saviors, others as Liberators – like Moses himself in 1,500 BC. -, but not in the first century, much less under those standards. Clearly there were more people called Jesus in ancient divided Israel, the Israelite conqueror Joshua himself, Moses' apprentice, receives the Hebrew name "Yeshua" (Salvation), like Jesus – then come the adaptations of the names to Latin, like Yeshua to Yehoshua or Jesua or Yesous.

The specific Greek term for Christ, however, was unique. It was the equivalent to the Hebrew Mashiaj, which we called Messiah, and which from the Hebrew language translated "Anointed".

Although, there are no two known Christs within this context and with these characteristics, so the Greek name of Jesus is the one that identifies him historically, in addition to giving a name to Christians. The New Encyclopædia Britannica (1995) states: "*These independent accounts show that in antiquity even opponents of Christianity did not doubt the historicity of Jesus, which began to be questioned, without any basis, in the late eighteenth century, throughout the 19th century and at the beginning of the 20th*». In the first century itself, the Samaritan historian Thallos also alluded in his writings to the darkness that occurred on the occasion of the death of Jesus, and tried to explain it as an eclipse of the sun. This part of his writings was later cited by the Roman historians Julio Africano and Phlegon Tralliano. In another case, the Talmud (a compendium of ancient rabbinic literature) contains several references to Jesus. They are inspired by an anti-Christian polemical attitude, which gives them a slanderous character.

These points may be of some use for a historical investigation about Jesus, not so much because of what they falsely affirm, but because of what they suppose: the historical existence of Jesus, his death sentence with the intervention of the Jewish religious authorities, his miracles (rejected as a product of magic): «*On the eve of the Easter festival, Jesus was hanged. Forty days before, the herald had proclaimed: "He is being led away to be stoned, for having practiced magic and having seduced Israel into apostasy. Whoever has something to say in his defense, come and say it." Since no one came forward to defend him, he was hanged on the eve of the Easter party.*» (Sanhedrin 43a). In the realm of Israel, there was a famous Jewish historian who spoke of John the Baptist (Jewish Antiquities 18:5:2), but he also let his knowledge about the public life of Jesus count.

This man is the best known of the extra-biblical witnesses who spoke of Jesus: the first-century Jewish historian Titus Flavius Josephus. Flavius Josephus referred to Jesus in two passages of his Antiquitates judaicae (Jewish Antiquities 18:3:3). The first of them is the famous Testimonium Flavianum about 93-94 AD. The received text reads as follows: «*At that time there existed a wise man, named Jesus, if it is lawful to call him a man; because he performed great miracles and was a teacher of those men who accept the truth with pleasure. It attracted many Jews and many Gentiles. It was the Christ. Rated on by the responsible princes among ours, Pilate sentenced him to crucifixion. Those who had loved him before did not stop doing so, because he appeared to them on the third day again alive: the prophets had announced this and a thousand other wonderful facts about him. From then until today there is a group of Christians that takes its name from him.*» Although it might seem that a Jewish witness was obviously supporting Jesus, it is not if we know the history and persecutions of the Jews. There was great dissension, because some accepted it and others did not.

Elsewhere Josephus wrote: "*Ananias was a soulless Sadducee. He cunningly summoned the Sanhedrin at the propitious moment. The procurator Festus had passed away. The successor, Albino, had not yet taken possession. He had the Sanhedrin judge James, the brother of Jesus, [called Christ] and some others. He accused them of breaking the law and handed them over to be stoned.* » (Jewish Antiquities, 20:9:1) Although the Antioch physician, the Syrian Loucas, who wrote the "gospel" that bears his name (Luke) and the Acts of the Apostles, was the best editor and notary of the first century, who compiled the best he could from the public life of Jesus and his disciples –he himself was an apprentice to Paul: « *Since many have already tried to put in order the history of things that have been very true among us, perhaps As those who from the beginning saw it with their eyes and were ministers of the word taught us, it has also seemed to me, <u>after having</u>*

diligently investigated all things from their origin, to write them to you in order, O most excellent Theophilus, so that you may know well the truth of the things in which you have been instructed.» (Doctor Luke 1:1-4)

The work of Luke was, a century later, recapitulated by Jerome in Rome, knowing that there was much literature on Jesus, which today has disappeared or remains hidden on purpose – taking into account all the manuscripts that have been burned, deteriorated and destroyed. throughout the centuries, such as the famous Gospel according to the Hebrews. In any case, Tiberio Cesar wrote to Pilate in very concise texts, regarding the unjust condemnation of Jesus: *«Because you had the audacity to condemn Jesus the Nazarene to death in a violent and totally iniquitous way and, even before pronounce condemnatory sentence, you put him in the hands of the insatiable and furious Jews; Inasmuch as, moreover, you had no compassion on this righteous man, but, after dyeing the cane and subjecting him to a horrible sentence and torment by scourging, you handed him over, through no fault of his own, to the torture of crucifixion, not before having accepted presents for his death...*

...because, finally, you manifested, yes, compassion with your lips, but you delivered with your heart to some lawless Jews; for all this, you yourself are going to be led before me, loaded with chains, so that you present your excuses and give an account of the life that you have given up to death without any reason. But woe to your harshness and shamelessness! Since this has reached my ears, I am suffering in my soul and I feel that my insides are crumbling. Well, a woman has come before me, who says she is a disciple of Him (she is Mary Magdalene, from whom, according to what she says, she cast out seven demons), and she testifies that Jesus worked marvelous cures, making the blind see and the lame walk, hearing the deaf, cleaning the lepers, and that all these cures were verified with his single word. How did you consent that he was crucified for no reason? Because if you didn't want to accept him as God,

you should at least have felt sorry for him as a doctor. Even the cunning report that has come to me from you, is demanding your punishment, since in it it is affirmed that He was superior to all the gods that we venerate.»

And in another letter, Tiberius explained to him: «*How was it to deliver him to death? Well, know that, just as you unjustly condemned him and ordered him to be killed, in the same way I am going to execute you with every right; and not only to you, but also to all your advisers and accomplices, from whom you received the bribe of death."* In this correspondence, Tiberio César will apply to Pontius Pilate the law of retaliation. It is strange that a Roman would apply Jewish law to a pagan like Pilate. The court that judged Jesus was more like a "tribunal of revenge", where Roman criminal law, procedural law and Hebrew law emanating from the Pentateuch and the Commandments of God given to Moses were violated. Of course, all this is irrefutable evidence, not only from writing, but from countless testimonies of the physical presence of Jesus throughout history, and where, exposing each testimony is already a matter of another type of investigation, just like those who claim, ridiculously, that the UFO issue is a phenomenon of collective hysteria.

FROM THE US LIBRARY of Congress.

Not many years ago, a certain person discovered in the United States Library of Congress in Washington some letters that recorded the historical existence of Jesus of Nazareth. Professor Alexander Backman translated them after receiving them from a friend of his, who had withdrawn them from said Library. He published these records in Ensanada (Baja California, Mexico) on April 27, 2011. These writings were personal letters between Tiberio Cesar and Pontius Pilate, also from Pilate and Herod. Some of these letters, kept in the custody of Rome, were even from civilians, soldiers, and other secular people. Among these mentioned words, we talk about the physical characteristics of Jesus and his way of being: *«A young man appeared in Galilee preaching with humble unction, a new law in the Name of God who had sent Him.»*

This particular letter added: *«One day I observed among a group of people a young man who was leaning over a tree, addressing the crowd in a calm manner. I was told it was Jesus. This I could easily have suspected. So great was the difference between Him and those who were listening to Him. His golden-colored hair and beard gave his appearance a heavenly aspect. He appeared to be about 30 years old. I had never seen a sweeter or more serene face."* This would have been Pontius Pilate's first personal experiences upon meeting Jesus, according to his own words. In another letter, a resident of Judea, at the beginning of the first century, under the reign of Tiberius Caesar, named Publius Lentrelus, spoke about Jesus, saying: "There lives in Judea at this time *a man of singular virtue whose name is Jesus Christ, whom the barbarians esteem as a prophet, but is loved and adore by his followers [as] the offspring of the immortal God. He calls the dead from the graves and heals all kinds of diseases with a single word or touch. [...] His entire speech, whether in word or deed, being eloquent and serious. [...] His manners are exceedingly pleasant, but He has often wept in the presence of men. He is temperate, modest and wise.»*

These last two texts can be found in E. Raymond Capt's book, "The Tomb of the Resurrection", and their sources first appeared in the writings of Saint Anselm of Canterbury, in the 11th century. Also found among these writings is a letter dated *« Calende de Abril 5th»*, from Pontius Pilate to the Roman Emperor Tiberius Caesar, trying to justify himself for the crime against Jesus. And likewise there are a couple of other records of other letters addressed to Caesar Augustus, also regarding Jesus' affairs in Judea, and others making Pilate almost a martyr who was condemned, although he tried to avoid the condemnation of Jesus. In fact, thanks to the writings, for example, of Flavio Josefo, it is attested that Jesus was condemned and murdered on April 3 and rose again on the 5th - more or less, since it is understood that he was in Hades for 3 days, not 2 -, around the year 33 of our era.

Another record is found in "The Archko Volume", where a chapter titled "Gamaliel's Interview" appears, where this wise Pharisee, Paul's teacher, declares interest in meeting Jesus, and someone answers him: "*If one day you meet him, you will know how to recognize it. Although he is not just more than a man, there is something about Him that distinguishes him from every other man [...] This Jew is convinced that He is the Messiah of the world... this same person who was born of a virgin in Bethlehem some twenty-six years ago.*» The said Archko Volume was translated into English by doctors McIntosh and Twyman of the Antiquarian Lodge, of Genoa (Italy), from the manuscripts in Constantinople and the records of the List of the Senate, taken, in turn, from the Vatican in Rome in 1896. Likewise, from other letters from a Syriac museum of the 6th-7th century that passed to the British Museum, Dr. Tischendorf drew material for his work "Apocalypse Apocryphae" (prolegg p. 56) – also with a Greek copy from a Museum of Paris -, words are read between Pontius Pilate and Herod, basically of anguish, for the unjust decision to consent to the murder of Jesus.

Among this material, a dialogue between Jesus, already resurrected, with Procla, Pilate's wife, several Roman soldiers and Pontius Pilate, who went to find Jesus in Galilee when they found out that he was alive: «*What is it? Do you believe in me? Procla, you should know that in the pact that God gave to the fathers, where it is said that all [the] world that had perished must live through my death, which you have seen. And now, you see that I live, whom you crucified. And I suffered many things, until the moment I was laid in the grave. But now, listen to me, and believe in my Father-God who is in me.*

*Because I released the bonds of death, and I broke the gates of Sheol (Hades); and **my coming will be after**."*

EVIDENCE THAT DEBUNKS *the Myth*

The main problem that exists, when it comes to wanting to talk about Jesus, is that they do not want to accept the historical versions that exist about him, for the mere fact that, whether it is true or not, they always try to associate him with religion. From the known story of Jesus in the Historia Patria de Israel – which Rome called the "Bible" - there are 3 narrations from 3 of his disciples: Matthew Levi, a former tax collector; John, the son of Zebedee, a fisherman; and Juan Marcos, who later accompanied Bernabé and Pablo. Then an important physician from Antioch appeared who recorded, not only the childhood of Jesus, but the life of Paul until he appeared before Caesar - including the most significant events of the early church and the apostles of Jesus. Jerome also took into account the sayings of Jesus documented by another disciple, Thomas Didymus, together with a letter from Barnabas - one of Jesus' disciples who was with him from the beginning - and a narrative "According to the Hebrews", which already today has disappeared, and which also spoke of the public life of Jesus in Judea.

It is ironic that religion itself is against Jesus. It is normal for Freemasonry and world power to be, but what about religion? Catholicism is an anti-biblical religion, by nature, and anti-Christian, although the shell makes it seem otherwise. As in Freemasonry, in Catholicism people from the lower ranks, up to the Cardinal, do not know that they serve Lucifer, whom the Freemasons call Iblis, an Arabic term, to call Satan. The Jews, who hold great world power, also try to deny Jesus, and if someone asks a lot about him, they say that he was a sorcerer or magician, as stated in the Babylonian document of the Sanhedrin. Then the Muslims, although they see him as a prophet, do not let anything, especially in Egypt -which represents symbolically, and is established, as a platform for the return of Jesus- there is something that legitimizes Christianity, as well as anything that can give reason to Israel to gain geographic territory. Even with everything, science has reached beyond the unimaginable.

The Chronovisor. Certain Vatican scientists made an intriguing discovery several decades ago. In front of Saint Mark's Square, and on the other side of the Grand Canal, in Venice, one of the most disconcerting mysteries -as well as ignored- of our days is locked up. On the island of San Giorgio, covered entirely by facilities run by Benedictine monks and the Giorgio Cini Foundation, dedicated to the care and education of orphaned children of fishermen, Father Pellegrino Ernetti hides from his past. Professor of "Prelolophony" (music from before the year one thousand) at the Benedetto Marcello Conservatory in Venice, Ernetti hides his research on the subject of Time. Father Ernetti did not want to give many explanations on this subject, of how he, helped by a large team of European scientists, had been designing - in the mid-forties - a machine capable of photographing the past. « *The principle is very simple: the visible and sound waves of the past are not destroyed. And they don't because they are energy. The greatness of our invention, which we call Chronovisor, lies in being able to recover that energy and recompose the scenes»*.

Hence the famous psychophonies that also study the voices that remain bouncing or static in one place for centuries. Ernetti made several hasty statements to the Italian press of the late 1940s. He claimed to have recomposed, in its original version, the officially disappeared work Thyestes, prepared by Fifth Ennius and performed in Rome around the year 169 AD. He also claimed to have obtained the original text of the Tables of the Law delivered by Jehovah to Moses on Mount Sinai, apart from other unique "photographs" obtained from the destruction of Sodom and Gomorrah, and other transcendental biblical episodes, including Jesus, who would have been photographed on Golgotha. The axis of his approach focuses on the admission of the existence of the ether, where each and every one of the external actions undertaken by human beings would be collected. According to Emetti, each one of us emits millions of waves throughout our lives, which remain trapped somewhere.

Later, thanks to the use of the appropriate instruments to access this stage of information and decode the waves that are being sought -in what, according to the French researcher Robert Charroux, a cathode oscillograph would be used to reconstruct the original emissions- You can access the images and sounds you want. The following interview was carried out by the researcher, and now director of the magazine Más Allá, Javier Sierra: *"But it's all over," says Father Ernetti. I already spoke. Pope Pius XII forbade us to divulge any details about this investigation, because the machine of the past is very dangerous. It can cut man's conscience of freedom, since with this device it will be possible to know what you have been doing this morning, where, when, how ...»*

Trojan Horse. Soon after, in the following decade, the 1950s, the US and the Soviet Union learned how to travel in time. This sounds totally absurd, but it is a statement from high-ranking military sources like Naval Intelligence officer Milton William Cooper, Colonel Phillip Corso, and more officers still alive today who belong to the huge Disclosure Project team led by Steven Greer. . The Disclosure Project is a mass of more than 200 people who belonged to the armed forces and politics, who now reveal the secrets of the United States government regarding wars, secret technology, space travel, the relationship with extraterrestrial civilizations and occult dealings for purposes of world domination (I deal with these topics in my works "Wandering Stars, the History of the UFO

Phenomenon", "Armageddon, E-5" and "Recognizing the End Time").

Studies in the field, such as the failure of the Philadelphia Experiment, were only the beginnings of a more thorough work of technological development that is 50 years ahead of what we officially know. JJ Benítez's book "Trojan Horse" may not be far from being the compilation of a story that occurred within the US military investigation to travel to the past, to the time of Jesus of Nazareth and keep track of his public life (personally I do not know all the details of these works, but I am aware of the thread of the subject, and especially I study, for a long time, conspiracy theories, among which are reversed technology and travel in time).

Blood of Jesus. Although this is beginning to border on logic, it is necessary, as a researcher, to evaluate all the probabilities and not rule anything out, as long as there are probabilities that they will help the reader and student to understand the importance of the historical life of Jesus, his message and his mission, in yesterday, today and tomorrow. Several decades ago, the American researcher Ron Wyatt, with the help of an Israeli soldier, inspected the area of Jerusalem where Golgotha (Jesus's crucifixion site) was located, and found, after much effort, a grotto containing petrified blood. Wyatt was famous for discovering Noah's Ark in Turkey, and getting the Turkish government to win a new visitor center at an insider's view of a 120 m long wooden boat lying on Ararat in the Ural Mountains. When Mr. Wyatt took this blood to some friends of his at the Hebrew University of Jerusalem, they studied it thoroughly.

Today, you can still find all the details of this investigation, which revealed something supernatural in the blood of Jesus, the one that Wyatt had found on Golgotha in Jerusalem: it had only 24 chromosomes. We know that the man gives 23 chromosomes and the woman another 23, which form an individual based on 46 chromosomes, where all the DNA information is. However, this man was supernatural, and had only ONE chromosome from the father, and, moreover, this blood, was ALIVE. Hopefully we will begin to open our eyes to the importance of Jesus, without the veil of religiosity, nor exclusively as a mere preacher of beautiful words that captivated the world.

THE CREATOR

The concept of Creator is the first thing that every tribe or ethnic group on Earth has considered. Even Plato philosophized about the existence of a promoter of things, which he called Demiurge. Ergo, who is that God? Usually one speaks of the same identity, which is the Father of all, Creator, God and Judge of existing things. He himself is questioned by humorists, critics, atheists, students and teachers, due to the social desire for a superhero to appear and solve all our problems. Not understanding who God is why people usually come to this sad point. As I mention in chapters 1 and 2 of "Reaching Deity", the concept of God is generic, and by definition is directed to Allah and Jehovah – possibly the same person -, but not to whom Jesus referred to as "the one" *Father*. What comes to be the team of Jehovah Elohim created a stage for Adam, Eve and their lineage, giving them autonomy over the planet, trusting in their good judgment to "subdue" it *and* submit it (Genesis 1:28. The Torah).

The means were within our reach, and, likewise, the support of Above. However, humanity became perverted and began to live selfishly, harming others, so that God decided to establish a day when *"he will punish the world for its wickedness"* (Prophet Isaiah 13:11), and he will judge *" each one according to be his work."* (Revelation of John 22:12) Everything that happens today in the world is the fault of Satan, indeed, but also of the lack of good will and altruism of men. That is why a famous saying goes: «*the only thing missing for evil to triumph is for good men to do nothing.*»

The Viennese scientific popularizer Thomas Vasek, supported by the work of the physicist Stephen D. Unwin, "Probabilities of the existence of God", establishes that God, and his non-being, start with 50%, which is the same as "not knowing anything » and divides the universe of study into five areas: 1) «the creation of the cosmos» concludes in favor of God, raising his score to 67%, since everything added is «somewhat more possible» that the cosmos was created since all science known prevents us from thinking that "something comes from nothing". 2) "the order of the cosmos" also plays in its favor, since a study of its physical conditions shows "a universe so improbable" that, if they changed one iota, it would collapse; this does not happen and leaves God very well with an 80% chance. Interestingly, in all other fields of human endeavour, we find that "design needs a designer".

Thus, design detection methodology is a prerequisite in many disciplines, including archaeology, anthropology, forensics, criminal science, jurisprudence, copyright law, reverse engineering, crypto-analysis, random number generation, and SETI (Search for Extraterrestrial Intelligence). In general, we find that "specific complexity" is a reliable indicator of the presence of intelligent design. Chance can explain the complexity, but not the specification. A random sequence of letters is complex, but it is not specific (meaningless). A Shakespearean sonnet is both complex and specific (has meaning). We cannot have a Shakespearean sonnet without Shakespeare. Not a creation without a Creator.

So, God, as Universal Father, is exempt from the affairs of the Earth, since he has only done his part as creator and monitor of living things. In his place, on Earth, a minister of his has been seen, known in the Scriptures as "The Ancient of Days", and who is explicitly mentioned in the book of the prophet Daniel, in the Apocalypse of John and in the Book of Enoch. It is this brilliant and majestic man, the one who represents God, and is known as

Jehovah. This man is the one who led the plan with Israel and who presides over the military forces that have to liberate the cosmos from tyranny, with the help of a prince and arch-strategist named Mijael (Miguel or Michael), who is the warlike defender of Israel. Although this sounds like science fiction, it is all put forward in the books of the proto-Hebrew scribe Enoch, who documented celestial affairs in times before the Deluge, and was one of the first to announce the coming of the Messiah. Why, then, is the Father not involved in this story?

Because he is "*holy, holy, holy*" and is exempt from where there is evil, since he has nothing to do with it, and he has also delegated his Creation into the hands of those whom He entrusts, so that in due time they will deliver it purified and Redeemed: «*For the creation was subjected to vanity, not of its own will, but because of him who subjected it in hope; because also the creation itself will be freed from the slavery of corruption, to the glorious freedom of the children of God.*» (Paul's Letter to the Romans 8:20-21)

Albert Einstein, the most famous scientist of the 20th century, when he was about to turn 71, reaching his last days of life in a hospital bed, even made notes on the sheets, where through mathematical calculations he tried to find a physics formula that summarized the entire existence of the universe and of life, which would be called TST (Theory Above Everything), or as he said: knowing the "Mind of God". Meanwhile, this obsession with wanting to prove the existence of God in the last days of his life, was leading him to ridicule in front of his physical scientific colleagues who did not accept his theory, since it collided with Quantum Mechanics.

This science foresees the unpredictability of the movements of subatomic particles. According to quantum theorists, everything that currently exists is the result of the existence of chance, that is to say, it is a mere case of nature; furthermore, according to these scientists, everything is summed up in a roll of dice, the law of the odds. What are the chances of rolling a six? However, the great genius did not believe in "perhaps-probably", at that moment he pronounced the phrase that would go down in history: "GOD DOES NOT PLAY DICE ". Behind each orbit of the planets, each star, each galaxy, each molecule, each atom, in short, each being, there is a SUPREME intelligence capable of moving and interacting with everything that exists, including life itself, which is called GOD.

ULTRA-CREATIONISM

Creationism is the philosophy that assumes that a deity, or deities, created everything that exists, both universally and at the level of nature (men, animals, plants and everything that surrounds them). Creationism is generalized as a Christian branch that intends to make war on Darwinism, however, creationism is applied to any other ideology, such as the Bahai religion or the ancient Nordic culture, where their gods as divine beings formed everything from nothingness or parts of a primal being. In Christianity, Judaism and Islam, creationism points out that the God "Yehovah", the only god, created everything in 6 days about 6,000 years ago, or at most 10,000 years. Among some of these created things would be the man "Adam", from where all human beings would come. This appreciation, like evolutionism, is a "point of view", rather than a correct concept drawn from sacred texts.

The studies of Felix G. Katchinsky maintain that Adam and Eve were not the first human beings to exist, although, nevertheless, they could have existed more than 6,000 years ago. Rabbi Felix G. Katchinsky, after more than 25 years of research into the Holy Hebrew Scriptures, is explicit in emphasizing that Adam and Eve probably lived in a world already inhabited by other human beings. For those of us who know Hebrew, it is clearer to argue that the 6 days of genesis would be 6 indeterminate periods of time in which the universe and living beings were not created, but rather a military establishment to restore order on planet Earth after a previous chaos.

His references are interesting, since they help to glimpse a way of seeing the beginning of time with a different, and at the same time innovative, prism. Now, a more accurate view of creationism would simply attribute the creation of life to the One Almighty God, not 6,000 years ago but billions of years ago, and not necessarily in our world in the first place. This more sane thinking would claim that modern creationism is in error and is a forced attempt to give value to myths and legends that are fundamental foundations in the tenets of many major religions. Objective creationism would say that the universe was created with everything that exists in it, from the most subtle and simple life forms to the most complex and macro-dimensional ones, as well as parallel universes, dimensions, laws of physics, radio waves. and of sound, air, water, fire, the minerals and gases that exist, etc.

But this event would have happened suddenly 15-21 billion years ago, and that Biblical and Koranic concepts – to cite these examples now – have been seriously misinterpreted or biasedly modified in the past for oppressive purposes.

This other segment or creationist vision would give more footing to extraterrestrial life and less to religiousness as the source or origin of many of the things done and the events that occurred in the prehistory of our world. Starting from the fact that at some point the ONE created the basis of life, its children, called today extraterrestrials and in the past angels, would have "recreated" life in the universe. Some of these would not necessarily be "angels" as soldiers-heralds, but as royalty or political leadership they would carry out the orders of the ONE, as can be understood in the beliefs of the UFO sects of Scientology and the Raelian, as well as the

Mormon point of view (although all these modern lines of thought are opposed to the teachings of Jesus).

These "foreigners" would be called Gods, from the Hebrew "Elohim", among them a faction uniquely called "Iehovah", while there would be a counterpart that, rejecting the parameters of the universe, would deceive the world by posing as "gods", as we see in the collection of mythologies and legends from around the globe. Not necessarily because a person is not an evolutionist should he be a creationist, nor should he be an evolutionist because he is not a creationist, the limits and "parties" are created by men themselves. The truth is not subject to appeals or prefabricated schools. This is slowly created with acquired knowledge, experience, human will, and the synthesis of cultural material that addresses the history of the world.

CREATIONISM COUNTERS Evolutionism

Until about 200 years ago, the Western world did not refute the idea that a Single God had created, by his divine power, everything in the cosmos, including all forms of life. When the ideas of Lamarck and Darwin appeared, the idea system of divine creation began to be seen as a worthless point of view. The archaic idea of divine creation took shape only for the heritage of mythologies, legends and religions as a fabulous point of view to give credit to the power of a god who, for many, as in the case of Marxism, would be a purely creation. human to justify themselves or what is the same, the idea or existence of God would have been manufactured by the masses themselves as a necessity and psychological comfort. Every year more and more college students believe that "*the scientific community is divided on evolution*" and that "*evolution is an unverified theory.*"

A survey carried out in 2001 by 'The Gallup' (an organization that has studied nature and human behavior for more than 70 years) shows that almost half of Americans believe in creationism. 45% of those surveyed think that God created the human being no more than 10,000 years ago, an idea very close to the creationist thesis. 37% of the other group believe that the finger of God, in any case, intervened at some point despite their belief in the evolution of species. Some of the supports that creationists use in the war between these two postulates (Creationism and Evolutionism) start from the fact that a few years ago Stephen Jay Gould and Niles Eldredge observed that the "great evolutionary lines" very often appear suddenly in the fossil record and They proposed that large-scale evolutionary change possibly unfolds gradually in some geological epochs, while it does so more rapidly in others. So, since this has not been seen in practice, this complicates the accepted hypothesis since these abrupt changes are even more difficult to support by chance over the years.

This main support gives rise, according to its defenders, to consider that God created everything overnight. The truth is that despite being correct the statement that all animal and plant groups in the fossil record appear suddenly from one day to the next and disappear just as quickly -which excludes a process of evolution- the claim is not admissible. creationist theory that maintains that God created everything about 6,000 years ago, or at most 10,000 years ago, since the fossil record is pragmatic in clarifying the millionaire antiquity of each animal and plant group. A study by Stuart Kauffman, a biochemist at the University of Pennsylvania, investigates how complex biological systems can self-organize from simple components. Some eminent creationists suggest that this "molecular gift" represents an alternative to natural selection, although evolutionists believe that the first forms of life likely arose on their own from the primordial soup.

The problem is that the temperature of this "original broth" and of the atmosphere made life impossible for any organism. Which means that no form of life, much less cellular, would have survived under such hellish conditions that reigned on planet Earth for billions of years. Founders of ID (Intelligent Design), a new and bolder branch of creationism, such as Michael J. Behe, a biochemist at the University of Pennsylvania, and other biologists, biochemists, chemists, physicists, philosophers, and historians wage war on Darwinism, and assume that the creator of man could even be of extraterrestrial origin. Behe proposes the idea of the irreducible complexity of natural systems. In other words, there are highly complex systems at the molecular level, such as the bacterial flagellum and the blood clotting mechanism, which cannot possibly have evolved on their own, which in themselves are evidence of design and not products of chance.

For his part, William A. Dembski, a mathematician at Baylor University, invokes that biodiversity is not explained by evolutionary chance and maintains that "the *action of creative intelligence leaves behind a characteristic sign or evidence that can be filtered and detect.*» (Source: "Muy Interesante" magazine No. 283 - December 2004) The reader will begin to see that everything we call science, rather than being the study of truth, has become a mere point of view subject to certain canons and parameters. from which you cannot leave, unless you want to be considered "ignorant" and lose the support of the system that prevails today. What we consider or believe to be science is not what it should be, and our seared mind only accepts what society considers acceptable, regardless of whether or not it is the truth.

«For Christ did not send me to baptize, but to preach the gospel; not with wisdom of words, so that the cross of Christ may not be made vain. Because the word of the cross is madness to those who are lost; but to those who are saved, that is, to us, it is the power of God. For it is written: I will destroy the wisdom of the wise, And I will cast away the understanding of the prudent. Where is the sage? Where is the scribe?

Where is the debater of this age? Has not God maddened the wisdom of the world? For since in the wisdom of God the world did not know God through wisdom, it pleased God to save the believers through the folly of preaching." (First Letter of the missionary Paul to the Corinthians 1:17-21)

THE HOLOGRAPHY OF THE *Universe*

Moving directly to another scenario, let's now talk about the origin of the universe and its laws. Is it possible that the universe and everything in it is the product of chance? If we look at the first of these conceptions, about the cosmos, we think, what force generated the Big Bang? Where did the matter that composed it come from? What existed before? These are questions that people do not always ask themselves, and the answer cannot be justified with the non-existence of God, otherwise it is like this. What force generated all the plasma in the universe? And all the more. What is the universe? How is it possible that everything came from nothing? A strange noise detected by the GEO600 (gravitational wave detector inaugurated in 2006 in Hannover, Germany) brought the researchers working on it upside down, until the physicist Craig Hogan, director of the Fermi National Accelerator Laboratory (Fermilab), in the USA USA, claimed that GEO600 had stumbled upon the fundamental boundary of space-time.

This is the point at which space-time stops behaving like the smooth continuum, described by Albert Einstein, to dissolve into "grains" (in much the same way that a photographic image can look grainy the closer you look at it). we observe). According to Hogan, *"it looks as if GEO600 has been hit by the microscopic quantum convulsions of space-time."* The physicist affirms that, if this is true, then the necessary evidence would have been found to affirm that we live in a gigantic cosmic hologram. The theory that we live in a hologram stems from an understanding of the nature of black holes, and while it may seem like an absurd theory, it has a fairly solid theoretical basis. The holograms of credit cards and bills are printed on two-dimensional plastic films. When light bounces off them, it recreates the appearance of a three-dimensional image.

In the 1990s, physicist Leonard Susskind and Nobel laureate Gerard 't Hooft suggested that the same principle could apply to the entire universe. If this is so, to begin with we should change our "mental chip" in terms of everything that surrounds us, and understand that "chance" could not create a three-dimensional hologram of billions of light years. Let's ponder, if the theory of cosmic evolution were correct and everything was the product of chance, we would find certain links, such as: What created the laws by which the universe is governed? Where did they come from: sound, light, radio waves, or dark matter? For example, if everything is a product of providence, would it also be the fact that our universe was not physical but "holographic"? The smallest unit that is used to understand the "life" pattern is the atom, and using it as a reference, let's take into account that it is mostly made up of an apparent "nothing".

In the same way, it is necessary to think that if there are other dimensions, if the universe is a three-dimensional projection - which some suggest is "artificial" -, and that contrary to what Albert Einstein once raised, the universe is increasingly moves faster, cosmic evolution does not answer the most elementary question: What then is the universe? Let's think that if the universe had originated by an outburst, the speed of movement of all matter in the cosmos would be less and less in a slowdown, and not greater. It is as if some unknown force is pushing them faster and faster.

But what is the universe? His own term is archaic and limited, since it includes a "single-verse" (universe) or "single version", since that is not what is being discovered today. What we believe to be the universe is more than 93,000 million light-years across, although that figure is always recalculated and falls short of new studies. If we think that we live in a 21 billion-year-old hologram, then what has happened during all that time, if we barely know anything about our recent past? This is said in some ufological and pseudo-epigraphic sources, and even scientific ones. It is also said that what we know of the cosmos encompasses 11 dimensions, although we hardly see and experience the Fourth Dimension. Each dimension or universe has its own laws, but who established them? *« There is much more than what we experience with our 5 senses»* (David Icke).

A UNIVERSE OF MULTIPLE Dimensions?

And what about the other dimensions that make up our universe? North American scientists detected in 2009 indications of the existence of other dimensions beyond the three known ones. Using data from the Amanda telescope, buried at the South Pole, they have been able to observe a dozen high-energy neutrino collisions with other elementary particles, thus obtaining the evidence of additional dimensions suggested by Superstring Theory. Analyzing the data provided, North American scientists have observed that neutrino collisions have an energy 10,000 times higher than that of the neutrinos emitted by our Sun with other elementary particles, thus obtaining evidence of the existence of other dimensions. Neutrinos are elementary particles of practically zero mass that are formed by nuclear reactions.

While the Sun and other cosmic phenomena produce low-energy neutrinos, high-energy neutrinos are produced by remote and extremely violent cosmic cataclysms, such as black holes, supernovae, and the Big Bang. Once formed by cosmic cataclysms, high-energy neutrinos travel close to the speed of light and never stop. As they have practically zero mass, they rarely collide with other particles, which allows them to travel in a straight line to the limits of the Universe, passing through stars, planets, magnetic fields and entire galaxies as if they did not really exist. Trillions of neutrinos traverse the Earth every nanosecond carrying with them crucial information about a series of cosmic phenomena and their origins. However, they are very difficult to detect, except when they collide with an atom. The collision disintegrates the nucleus of the atom and the neutrino is transformed into another particle called a muon.

On this we could mention black holes and how research on these has led to knowing more about the dimensions of our universe. Black Holes are massive objects which become so highly collapsible that their gravity pulls in anything in their proximity, like powerful vacuum cleaners in space. Certain very small black holes known as "extremal black holes" become massless at critical moments. How can this be possible in view of the extreme density? And how can they exert gravity without mass? Andrew Strominger theorized that the answer to its massless state lies in its extra-dimensionality. «*Strominger discovered that in six spatial dimensions, the mass of an extremal black hole is proportional to its surface area. As the surface shrinks, the mass will eventually become zero. The resolution worked out gave the existence of exactly six spatial dimensions* » (Hugh Ross, "Why I Believe In the Miracle of Divine Creation", from the apologetic anthology "Why I Am A Christian" by Norman L. Geisler and Paul K. Hoffman).

«*A theory solves two great dilemmas. This is what the theory tells us: The universe was created with ten rapidly expanding time-space dimensions. When the universe was just 10-43 seconds old, the motion when gravity separated from the strong-electroweak force, five of those ten dimensions ceased to expand. Today those six dimensions still remain a component of the universe, but they are just as crammed together as when the cosmos was only 10-43 seconds old* (editor's note:

"crammed together" is also called "Calabi-Yau space"). *Its cross sections are only 10 −33 centimeters, so small almost undetectable by direct measurement. Six sets of evidence indicate that this theory is correct. Perhaps the most convincing is that the chain theory yields, as an added bonus per product, all the equations of special and general relativity."* All this merges with quantum mechanics.

THE BIG BANG

The Big Bang theory maintains that the entire potential of the cosmos, some forty trillion galaxies, came out of a small point, smaller than a proton, which was an empty probabilistic reference frame of quantum mechanics, called a "scalar field". Furthermore, this empty point, a "false vacuum", contained not only the potential of "one universe" but "a hundred million universes". As has been so well described poetically by Gregg Easterbrook: *"You believe that when the Big Bang rang the universe expanded from a pinpoint to cosmological size in much less than a second, space itself hurling itself violently out of the torrent of physics." pure, the arc of the wave of the new cosmos moving at a speed of trillions of times more than the speed of light. You believe that this process unleashed such powerful distortions for an instant, the nascent universe was warped to a surreal degree. The extreme curvature caused normally rare "virtual particles" to materialize from a quantum world below cornucopia numbers, the matter of existence being "created virtually out of nothing," as Scientific American once put it."* (Gregg Easterbrook, "Science Sees The Light," the New Republic, October 12, 1998).

Hubble discovered a linear relationship between the distance to a remote galaxy and its "redshift," the apparent increase in the wavelength of emitted radiation. In the early 1900s, astronomers observed that light from distant galaxies moved toward the longer, or red, wavelength of the spectrum, interpreted as the galaxies moving rapidly away from each other. A blue shift would indicate that the galaxies were approaching each other. In 1965 radio astronomers detected faint radio waves wherever they pointed their radio telescopes. This confirmed the prediction in the 1940s by George Gamow, Ralph Alpher, and Robert Hermanque that if the universe were expanding from a singularity, then there must exist everywhere in the sky a faint background of radiation from that singularity event. a few degrees above absolute zero.

This subscription to the Big Bang model was further confirmed in 1922 and in 1933 by the COBE satellite which demonstrated that radiation from the cosmic plane fixed the spectrum profile of a perfect irradiator to a precision of better than 0.03% over full range. wavelength and as such has a trillion times more entropy (efficient in distributing energy) than a burning candle with an entropy of about 2.0. Only a very hot Big Bang can explain the gigantic entropy of the universe. This places permanently into oblivion the concept of the universe expanding and contracting cyclically, and likewise proves that the universe is only expanding. Verifiable predictions of light-element synthesis in the first minutes of the Big Bang, the universal abundance of Helium, remarkably constant from galaxy to galaxy, testify to a common cosmological origin.

Deuterium is destroyed in stars but not produced, and even traces of Deuterium are observed throughout the interstellar medium, as is the abundance of Lithium, which is also indicative of a common creation denominator. How can it be explained that the Big Bang arose and along with it the Laws of the Universe, 10 dimensions and, according to some, 22 universes parallel to it -creating a multiverse- being what we see only a huge three-dimensional cube of about 22,000 million years of antiguaty? There are too many coincidences not to attribute them to a superior being.

WHAT IS THE COSMOS?

The name "cosmos" comes from the Koiné (ancient Greek) word "cósmon" which means: "living universe", and its Hebrew and Aramaic equivalent is "olam", which means: "universe of the living". Both words are translated to English as "world". So the word commonly used by us as "world" is "the totality of created things", and it is misused to refer exclusively to our sphere. Therefore, if since antiquity the word cosmos is equivalent to world, and vice versa, the world is the universe of the living, a generic and plural term of "worlds", as it is regularly used in the Bible.

In other non-biblical ancient texts we see other interesting appreciations, such as what was said by the patriarch Abraham more than 3,000 years ago: « *Eli (My God) [is the] ONE, Universal Creator, has no beginning: it is eternal. Men are his children and he is his inheritance. "The worlds are infinite and man has to live in all those that exist today; but creation continues and does not end." "All the worlds communicate with each other in love and justice, and Eli is magnified in it". because I spoke with Adam and he looked like an angel; because I spoke with Eva and I saw her give birth to a savior and he is a son of Eli, who already lived in another world. "I am of the race of Adam and my children are of the race of Adam, who have to save the first race that populated the Earth; Because Adam and his family came with light and wisdom from Eli. "* (Secret Testament of Abraham).

Adam and his family came to save the original race that populated the Earth? Doesn't the Bible teach that Adam and Eve were the first? Do not Christians believe that the Earth is the center of the universe and men the crown of creation? How, then, does Abraham speak of other worlds and say that Seth -Adam's third son from where the human race is said to come from- already lived in another world? Are angels human beings? So where are they and where do they come from? How did Abraham know about the existence of other worlds? How could an archaic man have knowledge about the existence of other worlds and of humans living in those worlds and communicating with each other? Indeed, there is still a lot to bring to light.

HOW DID IT ALL COME About ?

According to the theory of Cosmic Evolution, of the evolution of stars from gas and of planets in their cooling process, we see the scientific answer to the formation of the Earth that says that our world was formed thanks to the gravitational attractive forces between the planets of our solar system and the rotational movement that continues in process throughout the universe since the Big Bang, as well as the high temperatures that "welded" them. According to this official postulate, the earth's core would have solidified into a mass of nickel and other hard minerals that would keep the external masses up to the earth's crust attracted towards the center of the planet. However, this theory is less likely in light of logic than the fact that our orb could be concave. At present scientists have discovered that our own interplanetary space is not empty.

For example, there are water molecules in space, the remnants of what are thought to have been clouds of ice crystals that seem to have engulfed stars in their early stages of development. This discovery supports the persistent Mesopotamian references to the waters of the Sun, which mixed with the waters of Tiamat. Basic molecules of living matter have also been found "floating" in interplanetary space - despite the fact that the temperature is so close to Absolute Zero -, shattering the belief that life can only exist within a certain range of atmospheres. or temperatures. Another fact that destroys the established theories and that try to justify everything without the divine hand, is the fact that there is "music" in space, such as the sound recorded for the first time near Jupiter. In addition, the idea that the only source of energy and heat available to living organisms is that emitted by the Sun has also been discarded. Thus, the Pioneer X spacecraft discovered that Jupiter, despite being much farther from the Sun than Earth, it was so warm that it must have had its own sources of energy and heat.

Scientists have also come to the unexpected conclusion that life not only evolved on the outer planets (Jupiter, Saturn, Uranus, Neptune), but in fact probably evolved there. These planets are made up of the lightest elements in the solar system. They are more like the composition of the universe in general, offering a profusion of hydrogen, helium, methane, ammonia, and probably neon and water vapor in their atmospheres—all the elements necessary for the production of organic molecules. In turn, modern science has come to the conclusion that life did not evolve on the terrestrial planets, with their heavy chemical components, but on the outer edges of the solar system.

Can Life be Generated in Any Circumstance?

Was it coincidence that only men evolved on Earth? If there is intelligent life on our world and it is nevertheless the only one we know of, couldn't the same conditions be provided on other planets? The discovery of life on other worlds is getting closer every day, according to the astronomical centers. NASA scientists have managed to detect basic molecules for biological activity on a gaseous extrasolar planet, in a move toward what the US space agency's Jet Propulsion Laboratory (JPL) has described as the goal of finding a cosmic body in which there may be living organisms. NASA indicated in a statement that the planet is not habitable, but has chemical activity, which, if it occurred on a rocky planet like Earth, could indicate the presence of life. " *This is the second planet outside our solar system where water, methane, and carbon dioxide, potentially important elements for biological processes on habitable planets, have been discovered,*" said JPL scientist Mark Swain. '*The detection of organic compounds on two exoplanets now raises the possibility that it is common to find similar bodies with molecules linked to life,*' he added.

The exoplanet was identified as HD 209458b, a gigantic gaseous body larger than Jupiter that orbits a star 150 light-years away in the constellation Pegasus. In December 2008, JPL scientists discovered carbon dioxide in another Jupiter-sized gaseous exoplanet identified as HD 189733b. Earlier observations by the Hubble and Spitzer space telescopes had revealed that there is also water vapor and methane on that planet.

The announcement of the discovery of organic molecules on the planet HD 209458b was made after an international group of researchers reported the detection of another 32 new exoplanets from the La Silla observatory in northern Chile. That figure brings the number of planets detected beyond the solar system to about 400. Swain and his science team made their find by using spectroscopic instruments to break down the light coming from the planet to identify its chemical components. The presence of the organic molecules was detected with Hubble's infrared camera and their quantity was measured by Spitzer's spectrometer, the statement said. "*This shows that we can detect molecules involved in the life process*" on planets beyond the solar system, Swain said. In addition, he added that scientists now have the possibility to compare the atmospheres of the two exoplanets to establish their differences and similarities.

Swain gave as an example that the amount of water and carbon dioxide is similar in both, but HD 209458b shows a higher abundance of methane than HD 189733b. «*The largest amount of methane can tell us something. Perhaps it means that there is something special about the formation of this exoplanet*," he added. At the beginning of 2009, NASA launched the Kepler probe into space, whose main mission is to search for rocky planets that could have similar characteristics to Earth. According to astronomers, it will take more than ten years on that mission before a planet can be found that could harbor signs of life like Earth's. Swain clarified that when the detection of organic compounds occurs, "*that won't necessarily mean that life exists on that planet because there are other ways to generate those molecules*." According to the scientist, exoplanets are too far from Earth to send probes to them and the only way to study them is through telescopes, whose spectroscopic systems are an important tool for determining their chemical composition and dynamics. (Source: http://ellaboratoriodedarwin.blogspot.com/ 2009_10_01_archive.html [1])

THE EARTH

1.	http://ellaboratoriodedarwin.blogspot.com/2009_10_01_archive.html

One of the many enigmas we come across is the appearance of life on Earth. Earth formed about 4,500,000,000 (4.5 giga-years) ago, and scientists believe that the simplest forms of life were already present a few hundred million years later. This is simply too early to get it. According to various indications, the oldest and simplest forms of life, more than 3,000 million years old, had molecules of biological origin, not of non-biological origin, as was believed and is still taught. This means that the life that existed on Earth so soon after the planet was born had to be, necessarily, the descendant of some previous life form, and not the result of the combination of chemical elements and lifeless gases. What all this suggests to bewildered scientists is that life, which could not easily have evolved on Earth, did not evolve on Earth.

In geological terms, the Phanerozoic eon is defined as the history of the planet from 570 million years ago to the present. It is the period of time that has been studied with more detail and attention, due to the richness of the geological remains and for building the classical fossil record. The entire time interval between the formation of the planet until the beginning of the Phanerozoic –from 4.55 Ga to 0.57 Ga before the present- has been called Precambrian or, more appropriately, Prephanerozoic. This period occupies 90% of the history of the planet, in turn divided into the Hadean eon (4.55 Ga to 3.9 Ga), the Archean (3.9 to 2.5) and the Proterozoic (2.5 Ga to 0.57 Ga). In the Archaic eon, the phenomena we call the "origin of life" must have occurred, the "transition from chemistry to biology". In fact, there are reasons to think that anaerobic microbial life – that is, developed in the absence of molecular oxygen – was flourishing 3.5 Ga ago.

During the Proterozoic eon, modern geological tectonics already occurred and the biological diversity, not morphological but biochemical, was very notable. Around 2 Ga (2 billion years) before present, the transition to an atmosphere with molecular oxygen occurred. This unique event in planetary history, it is taught, was possibly the catalyst for evolutionary innovations such as sex or multicellularity, and leads us, 1.5 Ga later, to the so-called Cambrian explosion, an incredible and sudden display of morphological diversity. , initiated with the Ediacaran and Burgess Shale faunas.

In the scientific journal Icarus (September 1973) he published an article stating that Nobel Prize Winner Francis Crick and Dr. Leslie Orgel advanced the theory that "life on Earth may have arisen from *tiny organisms of a distant planet*." They released their studies because of known discomfort among scientists about ongoing theories about the origins of life on Earth. Why is there only one genetic code for all terrestrial life? If life began in a primordial breeding ground, as most biologists believe, some variety of genetic codes must have developed. And in this order, why does molybdenum play a key role in enzymatic reactions that are essential for life, with molybdenum being such a rare chemical element on Earth? Why are elements so abundant on Earth, such as chromium or nickel, so unimportant in biochemical reactions?

But what was most unique about the theory put forward by these two scientists, Crick and Orgel, was not only that all life on Earth could have arisen from an organism from another planet, but that such "insemination" was deliberate - that intelligent beings from another planet will launch "the seed of life" from their planet to Earth in a spaceship, for the express purpose of beginning the chain of life on Earth. Curious statement for a Nobel Prize. If the theory of the formation of the planets is correct, that does not explain how the Moon is older than the Earth, if it was believed that it (the Moon) broke away from the Earth. Now, to this crazy theory of the appearance of life on Earth is added the fact that the current contact groups affirm that the majority of animal life, mainly dinosaurs, was brought from Mars to Earth and that the first Life forms on this globe were brought from The Constellation Cygnus. However, before we delve further into the nature of Earth and life on it, let's talk about the Moon.

The First Forms of Life

In science, as in many other areas of knowledge, it is very important to make it clear what we mean by each word we use. When talking about evolution we should clarify what kind of evolution we mean, since this hypothesis tries to hold on to many different parameters. In their eagerness to defend macroevolution (which is the transformation of matter into humans, for example) by natural processes, evolutionists often provide evidence for microevolution (which is variation within the limit of s species), thus its defenders voluntarily confuse microevolution and macroevolution, to such an extent that today one and the other concepts are spoken of together. The names of some organisms, such as Hallucigenia, show how taxonomists have marveled at such diversity - we know nothing about these organisms, since we can only grow less than 1% of existing microorganism species in the laboratory -.

All the theories on which we rely to assume the existence of creation are still vain speculations of the "changing mind" that we call science. Many times when scientists try to date the age of the universe based on 15 billion light years, they find that there are hundreds of stars much older than the age of the universe. At the moment of truth, we have not been to the Sun to find out how its hydrogen and helium are not consumed after millions of thermonuclear explosions in the middle of outer space with temperatures close to zero (-273.15 °C or -459 .67°F); neither do we know for sure anything about the origin of the Moon; we know only crumbs as to the origin of the universe; We take for granted concepts about the planets in our solar system, including Earth, that are incorrect.

How can we consider fable hypotheses as reliable stories? How can we take for granted theories that are just propaganda to deny God by an atheist world? The knowledge and approaches of everything that surrounds us must be based on objective and non-subjective structures and information with tendentious and manipulative purposes of the masses.

1,200 million years ago life on Earth succumbed to a meteorite. If life was just emerging, how did it survive? And what a coincidence that the little that was left managed to bring to forms more complex than a bacterium. If life on Earth was just emerging, we began to be contradicted by archaeological findings. If we accommodate the idea that life was brought here, we should reread some quotes from Sixto Paz Wells: «*Visitors came to our world from a planetary system in the Constellation of Cygnus, 6,000 light years from our Solar System. They planted spores in our world, about 3,000 million years ago, to change the acidity of the seas and make them alkaline, and thus modify the chemical conditions of the planet. This first humanity or extraterrestrial civilization is known as Antarctica, or the "Ancient Fathers".*» According to SP Wells, Pleiadian humans would bring patterns of life from Orion, which could explain the many representations of this constellation in prehistory.

Since 1990, Christopher Chyba - from the Institute for the Search for Extraterrestrial Intelligence - proposed that the water and gases in the Earth's atmosphere come from collisions with comets, meteorites, etc. that not only brought water and gases but also amino acids and other organic molecules. Evidence that this could have been the case is that the presence of kerogen, ethane and methane was detected in Halley's, Hale-Bopp and Hijakutake comets. Today this theory, called "panspermia" by the scientists who support it, points to the Orion Nebula as the possible origin of the first molecules on Earth.

Exclusive Design to sustain Life

1. The terrestrial temperature commutations are conserved within reasonable terms due to the practically circular orbit of the Earth around the Sun.

2. The Moon revolves around the Earth at a distance of 384,000 km causing the tides on the Earth. If the Moon were even 1/5th away from its location, the continents would be completely submerged twice a day.

3. Extreme temperatures are further moderated by water vapor and carbon dioxide in the atmosphere that cause the greenhouse effect.

4. Planet Earth is positioned at the correct distance from the Sun to receive exactly the right amount of heat to support life. The other planets in our solar system are either too close to the Sun or too far away to support life.

5. Any respectable permutation in the rate of rotation of the Earth would make life inadmissible. For example, if the Earth rotated 1/10 of its current rotation, all plant life would scorch from the heat during the day or freeze at night.

6. The Earth's atmosphere also serves to protect the planet from around 20 million meteorites that enter daily at speeds close to 48km/s. Without this protection the risk to life would be colossal.

7. The Earth has been particularly blessed with an abundant supply of water, a key substance of life due to its extraordinary and essential physical properties.

8. The thickness of the Earth's layer and the depth of the oceans seem to be carefully worked out. Enlargement in thickness or depth by a few centimeters would so strongly disrupt the absorption of free oxygen and carbon dioxide

that plant and animal existence would not exist.

9. The axis of our orb is pointing at 23.5º from perpendicular to the plane of its orbit. This adjusted tilt with the Earth's revolutions around the Sun causes the seasons of the year, which are decidedly essential for growing food supplies.

10. The atmosphere of our globe (ozone layer) operates as a defense shield against the lethal radiation of ultraviolet rays, preventing the annihilation of all life.

11. This planet has the right physical dimension and the exact mass to support life, allowing a careful balance between gravitational forces (basic to sustain water and the atmosphere) and atmospheric pressure.

12. Earth's magnetic field provides momentous protection from harmful cosmic radiation.

13. The two primary elements in our planetary atmosphere are nitrogen (78 percent) and oxygen (20 percent). This fine and critical distribution is fundamental to all forms of life.

«Such numerous, perfect and complex combinations of interrelated conditions and essential factors for delicate forms of life, unmistakably point towards intelligent design with purpose. To believe that such a complicated, carefully planned and balanced life support system is the result of mere change is truly absurd. Surely the honest and objective observer has no other recourse than to conclude that the earth-sun system has been carefully and intelligently designed by God for man." (Huse, Scott M., "The Collapse of Evolution"). Other sources: Riegle, DD, Creation or Evolution, "Creation or Evolution", Zondervan Publishing House, Grand Rapids, Michigan, 1971, pp.18-20. // Huse, Scott M. The Collapse of Evolution, "The Collapse of Evolution," Grand Rapids: Baker Books, 1997, third edition.

DOES SCIENCE ITSELF Contradict the Evolutionary Hypothesis?

The evolutionary hypothesis raises 6 primordial concepts, although many times it supports an evolutionary concept of ideas of another evolutionary concept. The Six Basic Concepts of Evolution:

- **Cosmic Evolution** (a big explosion produces hydrogen)
- **Chemical Evolution** (higher elements evolve)
- **Stellar Evolution** (stars and planets evolve from gas)
- **Organic Evolution** (life evolves from rocks)
- **Macroevolution** (changes between plant and animal species)
- **Microevolution** (changes within species)

Of all these, only Microevolution can be said to have been observed, and can be considered "science." It would be said that the other 5 are accepted by "faith". Can science and faith be reconciled? Can chance be an empirical test? Let's go step by step treating each topic with the magnifying glass. Let us also keep in mind that today Macroevolution is included with the study of Microevolution. The vast universe is an unknown neighbor from which we will always learn and never cease to amaze us. How did everything appear? This is a very common question among human beings, but what do we know about the universe? That is already another question that few ask themselves. Accustomed to being told what to think, what to believe, what to study or investigate, we are always guilty, like cattle or like a soldier, of being taken from here to there without questioning the facts. Points evolutionists want to ignore:

1. Thousands of species have not undergone any type of change despite the fact that millions of years have passed.
2. The very protozoan (microorganism) from which we are supposed to come, has not undergone any change in the last

6,000 years. In 6,000 years no modification or mutation has been seen in a monkey.

3. There are modern animals such as elephants, gorillas and tigers that are undoubtedly a degeneration of primitive species, if we base ourselves on the evolutionary point of view, which is no longer evolution but "regressive evolution".

4. Most species appear quite suddenly in the zoological annals, with nothing resembling a transitional organism ever to be found.

5. Although geologists estimate that there are more than 1 million animal species, evolutionists have been unable to find even the slightest acceptable trace of an intermediate state. To accept evolution they would have to find more than 3 million missing links.

6. Scientists agree that they have not found a subhuman life form to have lived on Earth.

Theory Violations

1. Belief in evolution is a violation of **the First Law of Thermodynamics**, the law of conservation of energy. It says: Matter is neither created nor destroyed, it only transforms. Nothing in the current economics of natural law can account for its own origins. The energy required for innovative evolution, for example, a fish growing legs to crawl out of a pond, violates the inviolable law of physics. The current structure of the universe is one of conservation. For example, the release of energy in an atomic reaction fission is not energy creation but a change of matter to energy. In the human being itself, it creates more bioelectricity than a 120 volt battery and more than 25,000 joules of body heat. Where does the "extra" energy come from to develop a superior creature?

2. The belief in evolution violates **the Second Law of Thermodynamics**, the law of the dissipation of energy. The energy

available for useful work in a functional system tends to dissipate, even though the total energy remains constant. Structured systems progress from a more orderly, more complex state to a less orderly, disorganized, and random state. This process is known as "entropy". Theoretically in a strange, limited and temporary situation a more orderly state could result. But according to this law, the tendency of all systems is towards deterioration. Evolution directly violates the Second Law of Thermodynamics. Evolutionists are aware of this and therefore it takes billions of years of constant violations of the Second Law of Thermodynamics. Statistically evolution is not only highly improbable but virtually impossible.

3. Evolution violates **the Law of Bio-Genesis** where life comes only from pre-existing life and is only perpetuated in its own kind. Bio = life; Genesis = generations. The belief in evolution is essentially a belief in "spontaneous generation" where in one scenario, life appears when lightning first strikes something dense and a living cell is somehow formed. Pasteur (1860), Spallanzani (1780), and Redi (1688) refuted that maggots can come from rotting meat, flies can come from banana peels, bees can come from dead cattle, etc., etc. When the deteriorated matter was sealed and pre-sterilized, there was no life or biological contamination.

Part III
GENETIC PERFECTION

*"The universe is a super-computer that interacts with,
not binary codes, but with the imagination."*
Michael Tsarion (Alternative Historian)

IMPOSSIBLE BY ITSELF

Our anthropocentric philosophy always hinders our vision of reality, and does not allow us to see further. This idea made us think about the theory that life had arisen on Earth from inanimate and dead matter. For the "miracle of life" it was only a matter of "stirring" a few drops of the original broth and with a little electrical sparking, highly complex proteins would be produced. Are miracles scientific? It is to be known that it is impossible for this supposed first primitive life to have been created on Earth by "own forces". Let's remember that the smallest unit we know of is the atom, which is subdivided into subatomic particles that are being investigated today to discover more about the microscopic world. Because it is not known what force holds the atoms together, it was speculated that there would be a unit called a "gluon" that held them together.

This hypothetical force that unites the atoms is nothing more than mere fantasy since nobody has ever seen them, nor has they measured them, that is, they do not exist. The electrons of the atom revolve around the nucleus hundreds of millions of times every millionth of a second and the nucleus of the atom consists of particles called neutrons and protons. Neutrons have no electrical charge and are therefore neutral, but protons have a positive charge. A law of electricity says: *"like charges repel each other"*. Since all the protons in the nucleus have a positive charge, they should repel each other and scatter into space. There must be some power that we do not know that holds the atoms together: *"in Him all things subsist"*, they are maintained. (Paul's Letter to the Colossians 1:17)

"Every organism is the way it is due to a long series of steps, all of them improbable." Carl Sagan (eminent scientist of the 20th century). Randomness in the universe is the common way of looking at things, to take the Creator out of the picture. Not taking into account such elementary points as the fact that a woman has XX chromosomes and a man an XY, which creates a similar being – if it were YY nothing would arise, it would be a hybrid. Another example is red blood cells, which do not contain DNA, and with the exception of sperm and egg, there are 46 chromosomes throughout the body. In the sperm and the egg there are 23, which makes it clear that it cannot be coincidental. To think that every detail of the universe, including life and procreation, is mere chance, is laughable. An overlapping point, more or less, in any of these elements would break the balance and make impossible a pure, natural or healthy result. It would be like saying that a woman is half pregnant (either she is or she is not).

Certainly, to begin to see things in a different way, we should go back to Darwin's days, but not to listen to him, but rather to his opposite, Gregorio Mendell, who already in 1860 began to talk about genetic transmission from parents to children. The same laws of genetics make evolution a pseudo-scientific hypothesis. In any case, let's see things for ourselves by sheer logic: the human body has 100 trillion cells (within each cell there are two pairs of genes). If the human genome were read at the rate of one word per second, it would take us - 8 hours a day - a century to read it. DNA, which looks like a very thin spring, has, wherever it is, the same chemical composition with regard to each of its proteins. DNA is made up of spirals where the information is that will determine what we will be. Each chromosome is made up of one cell, and all the chromosomes in all the cells in our bodies extend almost 2.6 billion kilometers.

The DNA strand is as long as the distance from Earth to the next galaxy. The DNA map is composed in such a way, and in such a language, that we can read and understand it. How is it possible? Let an atheist answer it. The sense of logic tells us that the perfection and the detailed finishing and elaboration of all existing things, basically as far as life is concerned, must be the result of a powerful, ingenious and superior mind in all rules - and is that perhaps? by mere chance different reproductive organs are created and working for something common? If Aristotle only began to know that there was hereditary participation of father and mother, how come the prophets of Israel, who are much older and were not indoctrinated in science, knew about genetics? Aristotle, the famous philosopher, began his speculations around the fourth century BC, long after Jeremiah (700 BC) or King David (900/1000 BC).

"*<u>Before I formed you in the womb I knew you,</u> and before you were born I sanctified you, I gave you as a prophet to the nations.*" (Prophet Jeremiah 1:5) How is it possible that without advanced science, microbiology was known? King Solomon's report wrote: «*Because* **you formed my bowels; You made me in my mother's womb.**" (Psalm 139:13) These are not the only Biblical references to knowledge that was not normal to have back then. Let's see it with the prophet Isaiah himself in 800 BC, when he said about Jehovah:

" *He is seated on the circle of the Earth* " (Prophet Isaiah 40:22). We know that the concept of the Earth as a sphere or round world was not notorious until the conquest of America, although in Egypt they had unofficially discovered that the Earth was round. The Bible speaks of other worlds, other civilizations, the Hollow Earth, the Last Ice Age, and other things that were impossible to know. Now, having taken a few minutes to figure it out for ourselves, let science itself answer a host of questions and paradigms we seemingly take for granted.

From the First Cell and Molecules

If all forms of life could be referred to an original form, this would be "the cell", since it is the smallest form of life, and it itself must have a "birth", it must have come from somewhere. Put then, we have to look for the composition of the cell and its micro-parts. Let us always keep in mind that the basis of all known life is the cell; the cell consists of macromolecules; macromolecule chains are atoms arranged in rows. These subatomic particles form the world of constant motion and diffuse radiation - an atom performs "10 to 23" pulsations per second. Thus we leave the material world to find the inconceivable, called God by some and spirit by others: "*There are no coincidences, we are all mathematical concepts.*" (Michael Tsarion, alternative historian). So now we have to analyze even the most ephemeral molecule of the cell, and this cell has to be reduced in its fundamental substance to a bunch of chemical elements.

But how were the chemical elements arranged in the sequence necessary to create the hereditary genetic substance or the cell itself? Similar questions led to the birth of "chemical evolution", and on this Manfred Eigen postulated that chemistry was subject to physical laws. It is known that physics has proven the existence of negative or positive electrical charges in each particle of matter. This law is also valid for molecules; according to their composition they should attract or repel each other. But in this the answers are just as slippery because the long macromolecular chains of the original broth were not only bound, but dissolved in the same way.

Many proteins are required for cell formation, the smallest conceivable protein consists of at least 239 molecules. Therefore, a protein molecule is a monster of different amino acids and enzymes that must all come together in a set order. Professor James F. Coppedge, longtime director of the Center for Biological Probability Research, in Northbridge, California, calculated the probability of such an ordering process at 1:10 to "23," that is, in a lottery. with a probability of hitting the prize of 1 against 10,000,000,000,000,000,000,000,000 0. And etc. If it is assumed that this cell, born from an impossible chance, was created in the conditions of the original broth and atmosphere, how did it survive? The atmosphere at that time consisted mainly of methane and ammonia, that is, oxygen would have had the effects of a deadly poison for the cell.

Thus, biology and prebiotic chemistry can only prove that a cell can only reproduce if it contains a complete, albeit modest, DNA program. This is sent from one cell to another, and this to the next, etc. Until forming a simple life such as a bacterium, for example. Evolution fails to explain the existence of even a "simple cell." The simplest unicellular organism has in its genes and chromosomes as much data as there are letters in the world's largest libraries: a trillion letters. There are hundreds of thousands of genes in each cell. Most life forms have such complex cells in perfect order that there is no room for them to arise by chance. There is no way that a random process could organize such a massive amount of data. The mathematical possibility that a human body is accidentally formed is the same as that an explosion in a printing press could form a dictionary, even if that explosion were experienced over and over again for centuries.

Sir Fred Hoyle, atheist, and creator of the "continuous-state" theory of the origin of the universe, believes that the chances of chance forming life on the planet are so small that they can be compared to chance that " *a Tornado going through a junkyard could assemble a Boeing 747 out of the materials found there.*" ("Hoyle on Evolution," Nature, Vol. 294, Nov. 12, 1981, p. 105). Hoyle and Chandra Wickramasinghe, a mathematical astronomer, calculated the possibility that life could have arisen spontaneously anywhere in a universe with a radius of 15 trillion light-years and at least 10 trillion years old. They found that the chance of this probability occurring is less than 1 in 1 with 30 zeros. Sir Fred Hoyle and Dr. Wickramasinghe have reluctantly come to the conclusion that life must have been created by a higher Intelligence (such as a kind of pantheistic intelligence that created the spores somehow in other parts of the universe and then were dragged to Earth), since it is extremely complex that they have arisen from natural processes.

Fred Hoyle makes another colorful comparison using a furry creature prized by evolutionists: "*No matter how large an environment one considers, life cannot have a random beginning. Even if we have troops of monkeys typing at random on a keyboard, the monkeys will not be able to produce Shakespeare's works for the practical reason that the entire observable universe is not large enough to contain the necessary hordes of monkeys, the required keyboards, and surely the waste baskets required for the deposition of wrong attempts. The same applies to living materials.*" (p. 148). Men will go out of their way to rationalize that there is no personal Designer of the universe who intelligently shaped all life. Evolution is a theory without scientific evidence to support it. It is an empty faith for those who do not want to believe in God and should be taught as just another religion; a religion inspired by Karl Marx to develop his theory of the struggle for classes and influenced by Adolf Hitler with his superior and evolved "Aryan super-man".

Many were sacrificed for his utopian and ruthless amoral vision. Evolution is a belief system that views the unborn fetus as an animal embryo that does not have the right to life and does not view it as God's creation. «*Anyone familiar with Rubik's cube [cube made up of smaller cubes with six different colors; the game consists in that all the cubes of each of the six faces have the same color] will admit that it is almost impossible for a blind man who moves the faces at random to solve the game. Now imagine 1050 blind men, each with a Rubik's cube with their mixed colors, and try to conceive of the probability that all of them would solve the game simultaneously. Then one would be likely to arrive, by random mixing, at just one of the many biopolymers [large molecules, such as DNA and RNA nucleic acids, or proteins] on which life depends. The notion that not only biopolymers, but also the operating program of a living cell, could be achieved by chance in a primordial organic "soup" here on Earth is patently utter nonsense.*" This quote comes from Sir Fred Hoyle, Honorary Research Professor at the University of Manchester and Cardiff University College.

Professor of Mathematics at the University of Cambridge. He is a well-known and highly respected scientist. In his opinion, the random development of life on Earth is "*extremely nonsense.*" Hoyle also says in another work devoted to biomolecules: "*... one must contemplate not only a single event to obtain an enzyme, but an immense number of attempts such as are supposed to have occurred in an organic soup early in the development of the enzyme.*" Land. *The problem is that there are about two thousand enzymes, and the probability of getting all of them in a random trial is only 1 in (10 20) 2,000 or 1 divided by 10 40,000, a ridiculously small probability that would hardly occur, even though the entire universe it was an organic soup.*" The least that can be said is that the chances of biopolymers and enzymes spontaneously forming and assembling themselves are, in Hoyle's opinion, "*ridiculously small.*"

RANDOM ODDS FOR LIFE

The bacterium itself, already as an example, is in itself, a finished life form with a certain function; must therefore be received its DNA genetic program from the first cell. How did the program for the construction of the entire bacterium come to the first bacterial cell? Where did the DNA of the first cell get the "order" and the "need" to build a bacterium? Furthermore, by means of what magic was one bacterium transformed into another, with "completely different functions"? For the probability that the simplest bacterium was produced by accidental modification, Professor Harold Morowitz, a physicist at Yale University, in the USA, calculated the following: 1:10 to 100,000,000,000... That's so many zeros to the right, which would not fit in this book. *« The probability that life would have originated by chance on one of 1,046 occasions is 10-255. The smallness of this number means that it is virtually impossible for life to have originated from a random association of molecules. The proposition that a living structure could have arisen in a single event through a random association of molecules must be rejected."* (Quastler, Henry - The Emergence of Biological Organization, New Haven and London, Yale University Press, 1964)

Not only could the cell (the smallest unit of life) not have appeared by chance in the primitive and uncontrolled conditions of Earth's early days, but not even the most advanced laboratories of the 20th century could have synthesized anything similar. Amino acids (the essential components of the proteins that make up the living cell), cannot by themselves build internal organs of the cell such as mitochondria, ribosomes, cell walls or endoplasmic reticulum, let alone a complete cell. For this reason, defending that life could have occurred in a first cell by mere chance is a figment of fantasy based entirely on imagination.

«Obtaining a cell by chance would require at least a hundred functional proteins that are simultaneously appreciated in one place.

This equates to one hundred simultaneous events, each with an independent probability that could hardly be more than 10 –20, giving a maximum combined probability of 10 –2000." (Denten, Michael. Evolution: A Theory in Crisis, Warwickshire, Burnett Books Limited, 1985) Harvard scientist John A. Ball, reflecting on the subject said: *"...Most evolutionists believe that it was generated (life) long ago, but perhaps never was... Perhaps the Earth was infected from somewhere else...."*

Regarding more mathematical calculations and probabilities for the origin of life, some even present the best formulas to objectively see the logic of thinking of evolution as a reality instead of as a way of seeing life: *« The The basic molecule of the genetic code is DNA. The more parts an organism has, the more complex it is. The simplest biological forms (although they lack the ability to reproduce on their own) are viruses. A virus has thousands of DNA or RNA nucleotides or "parts." For simplicity, let's invent a virus that has only one hundred parts. If there is only one correct way for the parts to be ordered, the probabilities of this happening in a single event are 1/100! This figure is read "one in a hundred factorial", and "hundred factorial" (100!) means 100 x 99 x 98 x 97... and so on up to... x 3 x 2 x 1. Let me give you an example of combination. If you had two blocks of wood, in how many ways could you arrange them in a straight line? The answer is 2!, that is, 2 x 1 = 2. If you had three blocks, the possible combinations would be 3!, or 3 x 2 x 1 = 6 combinations . If I had 4, they would be 4!, or 24 (4 x 3 x 2 x 1)...*

... The greater the number of parts, the greater the number of possible combinations. Technically, the "parts" of our virus could be arranged in a way other than a straight line, thereby greatly increasing the number of possible combinations. But we are being generous here. Now combining 100 in a straight line can be done in about 9.33 x 10 157 different ways. However, in the case of living beings, not any combination will do. Life supposes a delicate balance and therefore a

very precise combination of the component parts. Our problem now is to determine if 30 billion years is long enough for 100 parts to combine at a rate of 1036 combinations per second and result in life. The equation is simple. Thirty billion years is 3 x 10 10 years. In seconds, 3 x 10 10 years x 365 (days) x 24 (hours) x 60 (minutes) x 60 (seconds) this time corresponds to about 9.46 x 10 17 seconds...

... If this number of seconds is multiplied by the number of combinations that occur each second in our example, the result is 9.46 x 10 17 seconds x 10 36 combinations per second = 9.46 x 10 53 combinations, which we can round to to 1054 combinations. While this is a large number, it is extremely small compared to the 10,157 possible combinations. Subtracting 10 157 - 10 53 gives 9,999... x 10 156. Therefore, all the time in the world is not even close to enough for a single simple cell with 100 parts to spring into life. The probability does not differ practically from zero. If we were to observe cells with hundreds of other parts, restrict the time available (8 to 10 times less according to evolutionists themselves), and add some more realistic details regarding the number of combinations and environmental conditions, the odds against it would still be much higher.» (Source: http://www.maic.net/evolucion/matematica.htm [1])

⁓⦵⁓

WHAT WAS IT LIKE BACK then?

It is believed that on the primitive Earth there was still no living being. The model currently used by many scientists shows us a fairly hot crust, made up of primitive rock bathed by continuously boiling seas and in balance with clouds charged with rain and static electricity, which were discharged in the form of violent storms with lightning and lightning. As the temperature dropped, little by little, according to the opinion of science, other substances necessary for the eventual formation of the first molecules capable of

---

1. http://www.maic.net/evolucion/matematica.htm

self-reproduction were formed at random: formate, aspartate, lactate, glycine, ribose, adenine and glucose. Why can these molecules be formed? All of this, of course, was not the result of chance, but the intervention of a non-human force itself, adding also the "sowing" of life patterns, which was based not only on unimaginable chemical knowledge, but also on its intimate and secret relationship with geometry. For example, the use of the "tetrahedron".

It is interesting to know that both carbon, nitrogen and oxygen are fundamentally tetrahedrons, which somehow seek to assume that geometry in such a way that they will have the greatest stability when two electrons with opposite spin are found in the four vertices of the tetrahedron. This is important, since any other structure will be less stable and susceptible to react with other atoms. Since 1990, Christopher Chyba of the Institute for the Search for Extraterrestrial Intelligence proposed that the water and gases in the Earth's atmosphere come from collisions with comets, meteorites, etc. that not only brought water and gases but also amino acids and other organic molecules.

Biochemical evolution also talks about the formation of molecules in interstellar space and on Earth. AI Oparin and J. BS Haldane suggested this possibility, and HC Urey and S. L. Miller simulated it experimentally. But between the most complex mixture of organic molecules that we can imagine and the simplest living cell, there is a chasm that seems unbridgeable in the eyes of today's science. So what is life? the physicist R. Feynman replied: " *I don't know what life is but I know perfectly well when my dog is dead* ." Heraclitus of Ephesus said: " *We can only understand the essence of things when we know their origin and development* ."

METALS MAY BE OF EXTRATERRESTRIAL Origin

According to James Brenan, co-author of a study on the subject, he said: « *The extreme temperature at which the Earth's core formed more than 4,000 million years ago would have completely devastated any precious metals in the rocky crust and would have deposited them in the core .*» So the question for these scientists was; Why are there detectable concentrations of precious metals like platinum and rhodium in the rocky part of the Earth? " *Our results indicate that these could not have ended up there by any known internal process and that instead they must have been added as a 'rain' of extraterrestrial debris, such as comets and meteorites,"* explains the researcher . Geologists have long speculated that 4.5 billion years ago, Earth was a cold mass of rock mixed with iron metal that melted from the heat generated by the impact of planet-sized objects, allowing the iron to break apart. rock and form the Earth's core.

The scientists recreated the extreme pressure and temperatures of this process, subjected to a similar mix of temperatures in excess of 2,000 degrees Celsius, and measured the composition of the resulting rock and iron. Because the rock is stripped of metal in the process, scientists speculate that the same thing would have happened when the Earth formed and that some kind of external source, such as a shower of extraterrestrial material, contributed to the presence of some precious metals in the earth. the outer rock portion of present-day Earth. « *The notion of extraterrestrial rain could also explain another mystery, which is how the rocky portion of Earth came to have hydrogen, carbon and phosphorus, the essential components of life, which were probably lost during the violent beginning of the Earth* », concludes Brennan.

THE CELLULAR MYSTERY

The living cell, which still hides many secrets that we have not been able to reveal, is one of the main difficulties facing the theory of evolution. Another terrible dilemma from the point of view of evolution is the DNA molecule that is in the nucleus of the cell. It is a 3.5 billion unit coded system that contains all the details of life. DNA (Deoxyribonucleic Acid) is said to have been discovered in the 1940s and 1950s using X-ray crystallography, and it is a giant molecule with a superb plan and design that could not be the result of chance. For many years, Nobel Prize laureate Francis Crick believed in the theory of molecular evolution, but at one point had to admit that such a complex molecule could not have appeared spontaneously by chance as a result of an evolutionary process.

In 1980, two hemoglobin genes were sequenced. Although both encoded the same product, their nucleotide sequences differed by 0.8% if only substitutions of one amino acid for another were taken into account, and by 2.4% if amino acids present in one gene and absent in other were included in the comparison. other. Other genes sequenced later in other organisms led to the same conclusion: in the DNA sequence, the organisms may be heterozygous. The great variation revealed by these studies constitutes one of the foundations of the neutralist theory, another of the challenges to the synthetic theory. Its main expositor is Motoo Kimura, and in his opinion, most of the mutant genes are selectively neutral, that is to say, they do not have a selective advantage neither more nor less than the genes they replace; At the molecular level, most evolutionary changes are due to genetic drift from selectively equivalent mutant genes.

Genetic drift is the purely random change in gene frequencies, since any population consists of a finite number of individuals. The reason is the same as why it is possible to get "heads" more than 50 times when we flip a coin 100 times. Kimura stopped to think about the mutation probability that a mutant, appearing in a finite population, would itself show some selective advantage. That is, what is the probability that that gene will spread throughout the population? Kimura came to three findings:

1. For a given protein, the rate of substitution of one amino acid for another is approximately the same in many different phylogenetic lines.
2. These substitutions, instead of following a pattern, seemed to occur randomly.
3. The total rate of change in DNA was very high, on the order of one nucleotide base substitution every two years in a mammalian evolutionary line.

Regarding the variability within the species, it was found that most of the proteins were polymorphic, that is, they existed in different forms, and in many cases without visible phenotypic effects or a correlation with the environment. Thus, Kimura came to two conclusions:

1st Most of the nucleotide substitutions had to be the result of random fixation of neutral or near-neutral mutants, rather than the result of Darwinian selection.

2ª Many of the protein polymorphisms had to be selectively neutral or almost neutral and their persistence in the population would be due to the existing balance between the contribution of polymorphism by mutation and its random elimination.

The Russian evolutionary biologist, Alexander Oparin, claimed that the first living cell to arise from a common ancestor to all beings according to the theory of evolution could exist. In the 1930s, Oparin formulated a number of theories showing how the first living cell could form from matter by chance. However, all this failed and Oparin had to confess: " *Unfortunately, the origin of the cell remains a problem, really the darkest point of the whole theory of evolution.*" (Alexander Oparin, Origin of Life. p. 196) Jeffrey Bada, professor of biochemistry and defender of the theory of evolution, confessed in the Earth magazine of February 1998, in one of his articles on evolution: « *Today, while As we left the 20th century, we still face the biggest unresolved problem we had when we entered this century: How did life originate on Earth?* »

THE THEORY OF SPONTANEOUS Generation

This theory was popular in the late Middle Ages, and it argued that living things could easily arise from non-living things. It was common to think that mud worms and mealybugs arose spontaneously from damp places. For this reason some experiments were carried out, trying to prove this theory. For them, at first, the larvae in the rotten meat were "evidence" that life could be generated from non-living matter. But later it was understood that the larvae did not form spontaneously, but emerged from microscopic eggs deposited in the meat by flies. In Darwin's time it was very common to believe that life could originate from non-living matter, but a few years after the publication of "The Origin of Species", the famous French biologist Louis Pasteur scientifically refuted the theory of evolution. . Pasteur, after various studies and experiments, came to this remarkable conclusion: « *Can matter organize itself? No! There is currently no known instance under which microscope beings can be said to appear in the world without relatives they do not resemble.*" (Louis Pasteur, Origin of Life, p. 4-5)

PROBABILITIES OF LIFE from the Mathematical Point of View

Professor Bruno Vollmert, Professor of Chemical Engineering of Macromolecular Substances and at the time Director of the Institute for Polymers at the University of Karlsruhe, together with his team, spent decades investigating the creation of DNA in laboratories equipped with state-of-the-art equipment. The result of the investigation was overwhelming for all evolutionists: "*it is NOT possible that DNA could have been produced spontaneously.*" Vollmert says that a polymer chemist can neither be convinced nor be persuaded that the original mixtures arose by chance in chains of DNA-like macromolecules; the same is true, according to him, of the growth of DNA strands in the course of Earth's history, from one class of animals to the next higher. Vollmert's own words: "*Therefore, Darwinism is a way of seeing the world, an ideology, and not a scientifically proven theory... I believe that Darwinism represents a fatal error that owes its unparalleled success, ultimately, to once again, to an anthropocentric longing.*"

Since the formation of life has not been unequivocally elucidated, Professor Fred Hoyle, formerly Director of the Institute for Theoretical Astronomy at Cambridge, and Professor Nalin Chandra Wickramasinghe, Director of the Department of Applied Mathematics and Astronomy at Cardiff University, Wales -whom we mentioned earlier-, studied the possibilities of creating life, based on their mathematical knowledge. They wondered if enzymes could have arisen from an original terrestrial soup through chemical evolution. The conclusion of the two scientists was the following: « *We assume that the broth contains 20 biologically important amino acids in the same concentration. We advance the cautious proposal that 10 points above are decisive for correct biological functioning. More than "20 to 10" trials would be needed, in order to produce a single enzyme capable of functioning; and the probability of producing N quantity of such enzymes by chance amounts to "1:20 to 10N". Before N reached 100, the number of trials would have become greater than the number of atoms in all the stars in the entire Universe. We are therefore almost forced to conclude that life must be a cosmic phenomenon.*"

— ❦ —

THE IMPOSSIBLE GAP

New York University chemistry professor and DNA expert Robert Shapiro explains this belief of evolutionists and the materialistic dogma on which they are based: "Therefore, another evolutionary principle is needed to bridge the gap between *the mixtures of simple natural chemical elements and the first effective replicant. This principle has not yet been described in detail or demonstrated, but it is anticipated and is given names such as "chemical evolution" and "self-organization of matter."* According to Demirsoy, the possibility of the chance formation of Cytochrome-C, an essential protein for survival, is " *as improbable as the possibility of a monkey writing the history of humanity on a typewriter without making any mistakes."*

The origin of the mitochondria is the cell, and he openly accepts the explanation of "chance", even if it is "totally contrary to scientific thought": « *The heart of the problem is how the mitochondria acquired this distinctive character, because obtaining it by chance, even on the part of a cell, it requires incomprehensible extreme possibilities... The enzyme that provides respiration and works as a catalyst at each step and in a different way, composes the heart of the mechanism. A cell has to contain this entire enzymatic sequence, otherwise it is useless. Although this is contrary to biological thought, in order to avoid more dogmatic explanation or speculation, we have to accept, however grudgingly, that all respiration enzymes fully existed in the cell before the first cell. come into contact with oxygen.*

AMINO ACIDS FROM DIFFERENT Organisms

The amino acid sequence of a protein from two different organisms can be easily compared by aligning the two sequences and counting the number of positions in which the chains differ. They can be accurately quantified and provide an entirely new approach to measuring differences between species. As further work in this field became clear that each particular protein had a slightly different sequence in different species and that closely related species had closely related sequences. When hemoglobin sequences from different mammals, such as man and dog, were compared, the sequence divergence was around 20%, while when hemoglobin from two dissimilar species such as man and carp were compared, it was found that the divergence sequential was around 50%.

These comparisons make it possible to verify the hypotheses suggested by neo-Darwinian orthodoxy. For example, suppose bacteria have been around much longer than multicellular species, eg mammals. Suppose further that bacteria are more closely related to plants than to fish, amphibians, and mammals, in that order. If so, we should see evidence of these facts in the amino acid sequences of common proteins. That is, all the mentioned groups use Cytochrome-C, a protein used in energy production. The differences in this protein should be consistent with an evolutionary sequence. However, comparison of bacterial Cytochrome-C, the corresponding proteins in horse, pigeon, tuna, silkworm, wheat, and yeast, shows that the latter are all equidistant from that of the bacteria. The difference between the bacterium and the yeast is no less than between the bacterium and the mammal, or between any of the other classes.

Nor does it change if we choose other classes or different proteins. The traditional classes of organisms are identifiable through the typological hierarchy, and the relative distances between them are similar, regardless of the hypothetical evolutionary sequences. For example, Denton notes that amphibians are not found between fish and terrestrial vertebrates. Contrary to orthodox theory, amphibians are about the same distance from fish as reptiles and mammals. The really significant finding that comes to light when comparing the amino acid sequences of proteins is that it is impossible to arrange them in any kind of evolutionary series, and that the whole concept of evolution collapses because the pattern of diversity at the molecular level is conforms to a highly ordered hierarchical system. Each class is, at the molecular level, singular, isolated, and unrelated through intermediates. Furthermore, the accidental design adjustments that general evolutionism demands are logical disasters.

Random mutations due to radiation, copy errors, or other proposed sources rarely result in viable design adjustments, and never in more advanced perfect designs. Another important thing to keep in mind, amino acids prevent any transmutation of the species. The tissues of any part of the body take from the blood "the same corresponding amino acid", without exception. Each tissue produces a tissue of the same genus, this prevents any evolutionary process biologically speaking. The Turkish evolutionist professor Al de Misroy had no choice but to say the following on this subject: « *in fact, the probability of formation of a protein and a nucleic acid is smaller than we can calculate. What's more: the chances of a given protein chain appearing are so small that we can describe them as astronomical. The evolutionary theory that the first cell appeared by chance is completely irrational, comparable to saying that the Boeing 747 formed by chance." And that again at the cellular level."*

DOES AUTOPOIETICS GO against Evolution?

Another contradictory point in biochemistry, if one tries to support the hypothesis of molecular evolution, is about "autopoiesis". The concept of autopoiesis was introduced by F. Varela, from the Polytechnic School of Paris. An entity is autopoietic when, through chemical processes, it maintains and perpetuates its composition despite environmental disturbances. The mechanism of autopoiesis is metabolism, all the relationships of organic compounds catalyzed by enzymes, in aqueous phase or interphases, that occur inside living beings. The primary sources to start this metabolism are visible light and chemical energy. We know of no entity that is not autopoietic.

We also do not know of any entity that is not made of water and a complex diversity of organic compounds, including nucleic acids and proteins, making up a cell.

And in the fascinating book Mankind, Child of the Stars, Max H. Flint and Otto O. Binder detail a very important anomaly: "...molybdenum, a very rare metal, plays a *role important as a traceable element in the psychology of all creatures on Earth. It is surprising, therefore, that life so dependent on a rare metal arose in a world like ours, where molybdenum is so scarce ...*"

THE DNA MAP

The structure of the DNA molecule was discovered by scientists James Watson and Francis Crick in 1955. They discovered and demonstrated that life is much more complex than imagined. To confront evolutionists, Francis Crick, who received the Nobel Prize for this discovery, confessed that the structure of DNA could never have arisen by chance. Well, what is the genome? We have to say that this encompasses the total number of chromosomes in the body. Chromosomes contain approximately 80,000 genes, those responsible for heredity. The information contained in the genes has been decoded and allows science to know, through genetic tests, what diseases a person may suffer in their lifetime. The genome also makes it possible to treat diseases that were hitherto incurable, but knowing the code of a genome opens the doors for new ethical-moral conflicts, for example, the selection of embryos, or cloning beings to obtain stem cells and then discard them. This would threaten biological diversity and reinstate, among other things, the culture of a superior race, leaving others marginalized.

According to its detractors, those with a genetic disadvantage would be excluded from jobs, insurance companies, social security, etc. Similar to the discrimination that is said to exist in certain countries for certain jobs with women regarding pregnancy and children, for example. The genome (total number of chromosomes, that is, all the DNA: all the information of the DNA) of an organism includes in its genes all the information necessary for the elaboration of all the proteins required by the organism, those that determine the appearance, the functioning, the metabolism, the resistance to infections and other diseases, and also some of its procedures. In other words, it is the code that makes us who we are (ADAM). A gene is the physical, functional and fundamental unit of heredity, it is a sequence of nucleotides ordered and located in a special position on a chromosome. A gene contains the specific code for a functional product.

DNA is the molecule that contains the code for genetic information, which consists of a double helical strand held together by labile junctions between nucleotide base pairs. Nucleotides contain the bases adenine (A), guanine (G), cytosine (C), and thymine (T). The importance of fully understanding the genome lies in the fact that most diseases have a genetic component, both hereditary and those resulting from bodily responses to the environment. What power or force commanded these chains to intertwine? Once this was done, what intelligence ordered these strands to assemble correctly with the bases required to make up the DNA strands?

These two helically coiled strands that make up the DNA molecule are entwined around each other, like revolving ladders, whose sides made of sugar and phosphate molecules are connected by nitrogen bonds called "bases." There is a strict form of base union, thus forming pairs of adenine-thymine (AT) and cytosine-guanine (CG). Each daughter cell receives one old and one new strand. Each DNA molecule contains many genes, the physical and functional basis of heredity. The gene sends the specific sequence of base nucleotides with the information required for the construction of proteins that will provide the structural components to cells and tissues as well as enzymes for an essential biochemical reaction. In fact, viruses and bacteria are fundamental to life, they are not our enemies at all.

For example, when talking about junk DNA, some have suggested that if it is cleaned we can live up to 200 more years, but it has been discovered that the fundamental part of the genomes is precisely in this junk DNA. In the genome, the protein-coding fraction is 1.5% of the genes. A gene is the sum of gene sequences that are grouped specifically for each moment depending on the environment. For this reason, although we have 90% of the genes related to rats, that does not change anything, since we share 40% with a certain type of ecunet plant.

DNA and Chromosomes

The order or sequence of DNA specifies the exact genetic instruction required to create a particular organism with its own characteristics. Genome size is usually based on total base pairs. Only 10% of the genome includes the protein-coding sequence of genes. Interspersed with many genes are sequences with no coding function, the function of which was unknown until a few years ago. Chromosomes can be evidenced by light microscopy and when stained reveal patterns of light and dark bands with regional variations. Differences in size and band pattern allow the 24 chromosomes to be distinguished from one another, the analysis is called a "karyotype." Major chromosomal abnormalities include loss or extra copies, or major losses, fusions, or microscopically detectable translocations.

Thus, in "Down syndrome" a third copy of pair 21 or "trisomy 21" is detected. Let's analyze that something had to give us the possibility of having life, because without the correct order of chromosomes there would be no fertilization. Other changes are so subtle that they can only be detected by molecular analysis, they are called "mutations." Many mutations are involved in diseases such as "cystic fibrosis", sickle cell anemia, predispositions to certain cancers, or major psychiatric illnesses, among others. Every person has in their chromosomes in front of each paternal gene its corresponding maternal gene. When that pair of maternal-paternal genes (allemorphic group) are determinants of the same function or hereditary trait, the individual is said to be homozygous for that trait, on the contrary, it is said to be heterozygous. As an example we can mention that one gene transmits the hereditary trait of green eye color and the other the brown eye color.

They are heterocytogases for the eye color trait. If, in turn, one of these genes dominates the expression of the trait to the other opposite gene, it is said to be a dominant inherited gene, otherwise it is said to be recessive. Humans and chimpanzees are not only genetically different, but also differ considerably in the way they store DNA. DNA, the primordial blueprint of life, is packed into chromosomes, and all cells with a nucleus have a specific number of chromosomes. Many mistakenly believe that organisms, according to evolutionary theory, that share a common ancestor have the same number of chromosomes. This is not so because the number of chromosomes in living organisms varies considerably from one to another. For example, from 308 in the blackberry (Morus nigra) to 6 in animals such as the mosquito (Culex pipiens) or the nematode worm (Caenorhabditis elegans) (Sinnot, EW, LC Dunn, and T. Dobzhansky), Principles of Genetics (Columbus, OH: McGraw Hill, fifth edition, 1958)

Whether an organism has more or fewer chromosomes makes no difference. That is, complexity does not affect the number of chromosomes. The radiolarian (a simple protozoan) exceeds 800, but humans have only 46. Now, chimpanzees have 48 chromosomes. Exactly 2 more than humans, but this comparison cannot be used to create a relationship since by this rule man would be a relative of the hare (lepus europaeus), which has exactly the same number of chromosomes: 46; Or the Chinese muntjac (a small deer from the mountainous regions of Taiwan). The difference between the chromosomes of the man and the chimpanzee is the same as the man himself has with the hamster and the rabbit. If it were for the similarity of human (Homo sapiens sapiens) and chimpanzee (pan troglodytes) chromosomes, we must bear in mind that this (chimpanzee) has 48 chromosomes, just the same as a common potato (solanum tuberosum).

If the chromosome-bound DNA blueprint codes for only 46 chromosomes, how can evolutionary theory explain the loss of two entire chromosomes? The purpose of the chromosome is to reproduce constantly. If we deduce that this change in the number of chromosomes occurred through the evolutionary process, then this would imply that the DNA stored in the original number of chromosomes did not do its job well. If we take into account that each chromosome carries a certain number of genes, the fact that chromosomes are lost does not make physiological sense, and, most likely, it would be fatal for the new species. There is no reputable biologist to suggest that removing one or more chromosomes would produce a new species. The simple act of removing a single chromosome would potentially remove the DNA codes for millions of vital factors in the body.

On this, Eldon Gardner postulated: " *However, chromosome number is probably more constant than any other morphological feature that is available for species identification.*" (Gardner, Eldon J., Principles of Genetics (New York: John Wiley and Sons). 1968, p. 211). In short, there is no reason to look where there is none. Humans have always had 46 chromosomes and chimpanzees have 48. In other words, humans have always had 46 chromosomes, while chimpanzees have always had 48.

DOES HUMAN AND CHIMPANZEE DNA Indicate an Evolutionary Relationship?

DNA, the cell, and the complexity of life are the main drawbacks of evolutionary theory. In 1953, Francis Harry Compton Crick and James Dewey Watson proposed the double helical structure of DNA—the genetic material responsible for life. By demonstrating the molecular arrangement of four acid nucleotide bases (adenine, guanine, cytosine, and thymidine or thymine - A, G, C, and T) and how they combine, Watson and Crick opened the door to determining the genetic makeup of humans and animals. In 1962 these two scientists received the Nobel Prize in Physiology or Medicine for their discovery concerning the molecular structure of

DNA. Thirteen years after Watson and Crick received their Nobel Prize, the statement was made " *that the average human polypeptide is more than 99 percent identical to its chimpanzee counterpart .*" (King and Wilson, 1975, pp. 114-115).

However, this genetic similarity in proteins and nucleic acids left a major paradox—if our genetic material is so similar, why don't we look or behave like chimpanzees? King and Wilson recognized the legitimacy of this dilemma when they remarked: "*The molecular similarity between chimpanzees and humans is extraordinary because in anatomy and life they differ much more than other related species.* » (King, Mary-Claire and AC Wilson, "Evolution at Two Levels in Humans and Chimpanzees," Science. April 1975). Ergo, this seemed like the ideal answer that evolutionists wanted to support their hypothesis. A year after Watson and Crick's Nobel ceremony, chemist Emile Zuckerkandl observed that the protein sequence of hemoglobin in humans and in gorilla differed by only 1 in 287 amino acids.

Zuckerkandl noted: " *From the point of view of the structure of hemoglobin, it appears that the ape is just an abnormal human, or the man is an abnormal ape, and the two species actually form a continuous population.*" (Zuckerkandl, Emile, "Perspectives in Molecular Anthropology," Classification and Human Evolution, ed. SL Washburn (Chicago, IL: Aldine). 1963, p. 247). The molecular and genetic evidence only strengthened the evolutionary foundation for those who testified to our presumed primitive ancestor. Physiology professor Jared Diamond even titled one of his books The Third Chimpanzee, and there he considered the human species just another large mammal. All this pointed to the evolutionists being right: humans were more than 98% identical to chimpanzees.

However, after spending his life searching for evidence of evolution in molecular structures, biochemist Christian Schwabe was forced to admit: "Molecular *evolution is almost accepted as the superior method of paleontology for the discovery of relationships."* *evolutionary. As a molecular evolutionist, I should be excited. Instead, it seems puzzling that there are many exceptions to the orderly progression of species as determined by molecular homologies; in fact, there are so many that I think the exception, the quirks, carry the most important message."* (Schwabe, Christian, "On the Validity of Molecular Evolution," Trends in Biochemical Sciences. Jul 1986, p. 280). After the completion of the genomic mapping in 2003 the results began to become different. One of these resolutions would be that 1.5% of the human genome consists of genes, which code for proteins.

These genes are clustered in small regions that contain considerable amounts of "non-coding" or "junk" DNA between the clusters. The function of these non-coding regions is still completely unknown. Advancing science, the discoveries indicate that even if all human genes were different from those of a chimpanzee, the DNA could still be 98.5% similar if the "junk" DNA of humans and chimpanzees were identical. Jonathan Marks, an anthropologist at the University of North Carolina at Charlotte, has highlighted a certain relevant quirk about the problem that is often overlooked in this "similarity" line of thinking. Since DNA is a linear series of A, G, C, and T bases, there are only four possibilities at any given point in a DNA sequence. The laws of chance show that two random sequences of species that have no common ancestry will agree at one point in four. So even two unrelated DNA sequences will be 25% identical, not 0% identical (Marks, Jonathan, "98% Alike? (What Similarity to Apes Tells Us About Our Understanding of Genetics)," The Chronicle of Higher Education, November 12. May, 2000).

So a human and any other DNA-based terrestrial life form must be at least 25% the same. Marks went on to admit: " *Furthermore, genetic comparison is misleading because it ignores qualitative differences between genomes...So, even among such close relatives as humans and chimpanzees, we find that the chimpanzee genome is estimated to be about 10 percent hundred larger than human; that a human chromosome contains a fusion of two small chimpanzee chromosomes; and that the ends of each chimpanzee chromosome contain a DNA sequence that is not present in humans."* Let's look at a transcendental element, and that is that this difference of between 1-2% in DNA implies 80 million different nucleotides (3-4 billion nucleotides make up the human genome). Which is under the logical prism, a fairly wide difference.

THE REAL GENOMIC DIFFERENCES

Bert Thompson (Ph. D. of Apologetics Press, Inc. 2005) wrote: *"One of the ruins of previous molecular genetic studies has been the limit at which chimpanzees and humans could be exactly compared. Scientists would often use only 30 or 40 known protein or nucleic acid sequences, and from those would then extrapolate their results to the entire genome. However, today we have most of the sequences of the human genome, practically all of which have been made public. This allows scientists to compare every nucleotide base pair between humans and primates—something that wasn't possible before the human genome project. In January 2002, a study was published in which scientists had constructed and analyzed the comparative genomic map of a first generation human chimpanzee. This study compared the alignments of 77,461 final chimpanzee bacterial chromosome artificial (CBA) sequences with human genomic sequences."* Fujiyama and colleagues *" detected candidate positions, including two clusters on human chromosome 21, suggesting large, non-random regions of difference between the two genomes."* (Fujiyama, Asao, Hidemi Watanabe, et al., "Construction and Analysis of a Human-Chimpanzee Comparative Clone Map," Science, January 4,

2002). In other words, this comparison revealed huge differences between the chimpanzee and human genomes.

The authors found that only 48.6% of the entire human genome matched chimpanzee nucleotide sequences. «*Only 4.8% of the human "Y" chromosome could correspond to the chimpanzee sequences.*» Add, Bert Thompson. Conducting this study compared alignments of 77,461 chimpanzee sequences with human genomic sequences obtained from a public database. Of these, 36,940 final sequences were unable to be mapped into the human genome. It was speculated that nearly 15,000 of those sequences that did not match human sequences "corresponded to human sequence-free regions or were from chimpanzee regions that have diverged substantially from humans or did not match for other unknown reasons" (295: 132).

Although the authors noted that the quality and usefulness of the map should "increasingly improve as sequencing of the human genome progresses" (295:134), the data already supports what creationists have said for many years—the The 98-99% figure representing DNA similarity is extremely misleading. (Sources: Fujiyama, Asao, Hidemi Watanabe, et al., "Construction and Analysis of a Human-Chimpanzee Comparative Clone Map," Science, January 4, 2002; Bert Thompson, Ph.D. Apologetics Press. www.apologeticspress.org) In the wake of another study, Barbulescu and his colleagues similarly discovered another significant difference in the genomes of primates and humans. The authors wrote: "*These observations provide very strong evidence that, for some fraction of the genome, chimpanzees, bonobos, and gorillas are more closely related to each other than they are to humans.* » (Barbulescu, Madalina, Geoffrey Turner, Mei Su, Rachel Kim, Michael I. Jensen-Seaman, Amos S. Deinard, Kenneth K. Kidd, and Jack Lentz, "A HERV-K Provirus in Chimpanzees, Bonobos, and Gorillas, but not Humans," Current Biology, 2001).

This demonstrates the mistake of evolutionists in stating that chimpanzees are genetically closer to humans than gorillas. Another study using an interspecies "figurative difference analysis" (ADF) between humans and gorillas revealed gorilla-specific DNA sequences (Toder, RF Grutzner, T. Haaf, and E. Bausch, "Species-Specific Evolution of Repeated DNA sequences in Great Apes," Chromosome Research, 2001). That is, the sequences found in gorillas, but not in humans " *may represent an ancient sequence that is lost in other species, such as man and the orangutan, or, more likely, they represent recent sequences that evolved or originated specifically." in the gorilla genome.*

As early as 1998, a structural difference between the human cell surface and that of the ape was manifested. After studying tissue and blood samples from great apes and 60 humans from various ethnic groups, Muchmore and his colleagues discovered that human cells lack a particular form of sialic acid (a type of sugar) found in mammals (Muchmore, Elaine A., Sandra Diaz, and Ajit Varki, "A Structural Difference Between the Cell Surfaces of Humans and the Great Apes," American Journal of Physical Anthropology, October 1998). « *This sialic acid molecule is found on the surface of every cell in the body, and is thought to perform multiple cellular tasks. This seemingly minuscule difference can have far-reaching effects, and may explain why surgeons couldn't transplant chimpanzee organs into humans during the 1960s. identical to us", simply because of a genetic coincidence."* states Bert Thompson, Ph. D. of Apologetics Press.

PARALLELS AND DIVERGENCES

The complete genome of the nematode (Caenorhabditis elegans) has also been sequenced as a side study to the human genome project. Of the 5,000 best known human genes, 75% of them have combined with those of the worm ("A Tiny Worm

Challenges Evolution". Video available at: www.cs.unc.edu/~plaisted/ce/worm.html). Homology (similarity) does not prove common ancestry, or are we 75% identical to the nematode worm? " *The fact that living creatures share some genes with humans does not mean that there is linear ancestry*." (Bert Thompson, Ph.D. of Apologetics Press).

The biologist John Randall admitted this when he wrote: "*Old evolution textbooks emphasize the idea of homology, pointing out the obvious similarities between the skeletons of different animal limbs. Thus, the design of the limb "pentadactyl" (from the Greek language: "five bones") is found on the arm of man, the wing of the bird, and the fin of the whale—and this is considered to indicate their common origin. If these various structures were transmitted by the same pair of genes varied from time to time by mutations and changed by environmental selection, the theory would make good sense. Unfortunately, this is not the case. Homologous organs are now known to be produced by entirely different gene complexes in different species. The concept of homology in terms of similar genes that have been passed on from common ancestry has failed ...*' (Fix, William R., The Bone Peddlers: Selling Evolution (New York: Macmillan) p. 189).

Elaine Morgan commented on the difference between humans and chimpanzees: "*Considering the intimate genetic relationship that has been established by comparison of the biochemical properties of blood proteins, protein structure, and DNA and immune responses, the differences between a man and a chimpanzee are more striking than the similarities. These include structural differences in the skeleton, muscles, skin, and brain; differences in posture associated with a singular method of locomotion; differences in social organization; and finally the acquisition of speech and manipulation, along with the dramatic growth in intellectual capacity that has led scientists to name their own species Homo sapiens sapiens—wise wise man. During the period that these remarkable evolutionary changes were taking place, other closely related ape-like species were changing very slowly, and with far less remarkable results ...*

...It is hard to resist the conclusion that something had to have happened to the ancestors of Homo sapiens that did not happen to the ancestors of gorillas and chimpanzees." (Morgan, Elaine, The Aquatic Ape: A Theory of Human Evolution (London: Souvenir Press) 1989, pp. 17-18) Other Sources: Shouse, Ben, "Revisiting the Numbers: Human Genes and Whales," Science, 22 February 2002; mainstay of argument on this topic about chromosomes: Bert Thompson, Apologetics Press (230 Landmark Drive, Montgomery, Alabama, USA, http://www.apologeticspress.org) Genome study shows that our DNA is similar to a computer database or a master computer. Without the specific sequences it doesn't work, and without a flow of energy nothing interacts. Who or what is that invisible force that gives life to all existing things from the microscopic to the macroscopic world? If the genome shows us the unique and complex properties of our DNA.

How can we think that there are external alterations, whether physical, chemical or biochemical, that modify the DNA information without having also been created by a superior intelligence for said function? The inexplicable, «... *the missing link in the history of man's evolution, was the controlled artificial manipulation practiced in one of our most remote pasts.* » (Erick von Däniken. Swiss writer of several "Best Sellers").

MUTATION?

Mutations are defined as substitutions or breaks that take place in the DNA molecule, which is found in the cell nucleus of a living organism and contains all the genetic information. These substitutions or ruptures are the result of external effects such as chemical action or radiation. Each mutation is an "accident" that damages the nucleotides that make up the DNA or changes their location. Most of the time it causes so much damage and modification that the cell cannot repair it. Mutation, to which evolutionists often allude, is not a magic wand that transforms living organs into a more perfect and advanced form, that only happens in movies and fantasy tales. The direct effect of mutations is entirely harmful.

The changes effected by mutations can resemble only those experienced by the people of Hiroshima, Nagasaki (Japan) and Chernobyl (USSR), that is, death, disability and the abortion of nature... The reason for this is very simple: DNA has a very complex structure and disruptive effects can only cause damage to that structure. Says BG Ranganathan: « *Mutations are small, random and harmful. They occur rarely and are most likely ineffective. These four characteristics of mutations imply that they cannot lead to evolutionary development. A chance change in a watch cannot improve it. Most likely, it will damage it or, at best, it will not affect it. An earthquake does not improve the city it hits, but causes its destruction.*" (BG Ranganathan, Origins?, Pennsylvania: The Banner Of Truth Trust, 1988)

Scientists from the University of Liverpool discovered that the interaction between two species battling each other is more important in the evolutionary process than the interaction of the organism with its environment. The discovery was published in the February 24 (2010), issue of the prestigious journal Nature. Dr. Steve Paterson and his colleagues at the University of Liverpool designed an experiment to see which of the two theories actually guides biological evolution. Viruses and bacteria of two different species were placed in an environment. Scientists looked at how viruses attacked bacteria, how bacteria evolved defenses, how viruses tried new methods of attack, how bacteria reacted, and so on.

After hundreds of generations, it was discovered that both species evolved at an accelerated pace compared to a controlled population of viruses and bacteria that did not compete in this way. This experiment suggests that the environment is not the predominant agent in evolution, but the competition for survival and self-defense between species, and even so it does not imply that there is an external mutant or chemical alteration to modify the genetic information of the bacterium. but some unknown force that forces her to become immunized. But it doesn't stop being a bacterium to become an ant. Dr Lee Spetner is an expert in genetics and states in his book that " *Every mutation that has been studied at a molecular level reduces genetic information and does not increase it.*" (Not by Chance, The Judaica Press, New York, p.138, 1997)

DOES LIFE HAVE EXTRATERRESTRIAL Origin?

A team of scientists has obtained a new hint that some key ingredients for the necessary chemical development that led to the rise of RNA and DNA could have come from outside our planet. The research has been carried out by experts from Imperial College London, NASA, the University of Maryland in Baltimore, the Carnegie Institution of Washington, the Open University Space and Planetary Sciences Research Institute in Great Britain, Radboud University in Nijmegen (Netherlands), and the Laboratory of Astrobiology of the Leiden Institute of Chemistry (Netherlands). The 1969 meteorite that fell near Murchison, Australia, is famous for the large number of organic compounds found on it, including nucleobases, which are precursors to the constituent molecules of RNA and DNA.

This led the scientific community to consider that the fall of meteorites such as this one in an archaic period could provide the Earth with the key ingredients for the emergence of life, and that therefore the forms of life in our world would have a partially originated alien. However, there was doubt about the origin of the nucleobases present in the meteorite, since it could have been contaminated with terrestrial material, and therefore those detected in it would not have an extraterrestrial origin but a completely terrestrial one. Now, the authors of the new study have succeeded in isolating xanthine and uracil from the meteorite, and subjecting them to isotopic analysis. The ratio between different carbon isotopes is an unmistakable fingerprint of the provenance of organic molecules. Those of extraterrestrial origin have higher abundances of carbon-13 compared to carbon-12.

The result of the analysis shows that the nucleobases present in the Murchison meteorite come from outside our planet. This implies that the hypothesis of the extraterrestrial origin of life in our world is certainly plausible. " *We believe that the earliest forms of life were able to adopt nucleobases from meteorite fragments for use in the genetic code that enabled them to pass on beneficial traits to subsequent generations,*" says lead study author Zita Martins of Imperial College. From london. On the other hand, between 3.8 and 4.5 billion years ago, vast amounts of rocks such as the Murchison fall in 1969, reached the Earth's surface from space. That meteorite bombardment, which left numerous craters on stars in our solar system, coincides with the time when, according to all indications, life arose on Earth. (Sources: amazings.com/ciencia and masalladelaciencia.es)

EVOLUTIONARY DEVELOPMENTAL Biology

Evolutionary developmental biology (or informally "evo-devo", from English evolutionary developmental biology) is a field of biology that compares the development process of different organisms in order to determine their phylogenetic relationships.

Thus, evolution is defined as change in development processes. The approach adopted by evo-devo is multidisciplinary, bringing together disciplines such as developmental biology (including developmental genetics), evolutionary genetics, systematics, morphology, comparative anatomy or paleontology. During the 1980s and 1990s, a large amount of comparative data on the molecular sequence of different types of organisms was collected and the molecular basis of the developmental mechanisms encoded by such genes began to be understood in detail. results provided by the new genetics of development.

Evolutionary developmental biology began to become a discipline thanks to the availability of such data. Among the most surprising and probably most counterintuitive results of evolutionary developmental biology research is the fact that the diversity of organismal body plans and morphology across many phyla is not necessarily reflected in one diversity at the level of gene sequences involved in the regulation of development. In fact, as Gerhart and Kirschner (1997) point out, we find ourselves with an apparent paradox: *"where we expect to find variation, we find conservation, the absence of change."*

The totality of things made, from the Big Bang to nothing to everything, from subatomic particles to macromolecules and complex and intelligent life forms must have been created by someone higher. This is the most consistent. Ergo, if this is so, and the theory is wrong in every way, where did we come from if not by an evolutionary process from ephemeral particles? They have taught us that there are links of man that support the theory of evolution, however, these statements are totally false as we will show later. So where do all these hominid species like Neanderthals, Homo Habilis, Homo Erectus, etc. come from? Scientists were based on the fact that they had evolved towards each other, but evidence and studies show that they were not as cute as they have been painted to us. We could say that there are only two groups: monkeys and men.

WHAT POWER CREATED *Love in the DNA?*

Fear has a long, slow vibration frequency, while love has a very fast, high frequency. To show that vibrations are the very basis of our existence, Hans Jenny developed what are known as "cymatics" in the 1940s, demonstrating that when sound vibrations are transmitted through a communication medium, the set generates a model to follow when the frequency increases. Fear develops a more complex pattern, and this is precisely what is happening to our planet and to humanity. There are 64 possible combinations of amino acids in our DNA structure, made from 4 elements: carbon, oxygen, hydrogen, and nitrogen. According to logic, we should have all 64 active combinations within our DNA structure. However, currently there are only 20 active codes.

According to studies by Gregg Branden: *"... of these 64 possibilities, it seems that only 20 of these codes are currently becoming our DNA. The 20 amino acids. There is a switch that turns on and off the possibility of these codes appearing, and this switch is what we call 'emotion'. It is the first time we have seen emotion patterns directly and physically linked to human genetic material. Well, fear is a long, slow wave, so this long, slow wave of fear touches relatively few places in our DNA, so a person who lives permanently in a state of fear limits their 'antenna' of the spectrum that have at your disposal. Considering that a person lives according to the pattern of love, this is love, and you can see that it has a higher frequency and a shorter wavelength, we have many more potential sites for genetic coding along that pattern. This information is amazing. It's the first time we've had a strong digital link between emotion and genetics,"* adds Braden.

This is important to understand why another researcher named Vladimir Poponin made measurements of tiny particles of light, called photons, inside a vacuum tube, noting that the photons scattered as expected. When he introduced a DNA sample into the vacuum tube and measured its dispersion again, it was found that the photons added along the axis of the DNA strand subsequently, as the DNA sample was removed the photons kept aligned according to the same shape of the DNA. Even without this one being present. This is what is known as "ghost DNA testing". Science has here an important bridge to unite the physical and the ethereal or spiritual. Emotions directly affect the structure of our DNA, which shapes the physical world we experience every day.

In the documentary Esoteric Agenda we can see this more concisely: « *Therefore, the messages left by the ancients that have already been explained, are something more than prophecies about a government or New World Order. We now understand why studying the heavenly bodies was so important. The rotation and orbit of everything that makes up our universe works like a clock, marking changes in route and transitions. This helped the ancients to understand the changes in the celestial bodies were a mirror that reflected the changes of all existence. December 21, 2012 is simply a natural transition from one energy form to the next. The transcendental evolution of man. This date is what is known as zero point. Our Sun, as well as our planet Earth, are decreasing its magnetic field, and performing its rotation, at the same time that its resonant cavity base frequency, also known as "Schumann resonance" increases as the Fibonacci sequence is predicted.* »

OTHER EXPERIMENTS

At the cellular level, our bodies respond to electromagnetic pulses, which the ancients called the "sacred circle." The cells receive this pulse from the brain, which receives it from the heart, which receives the pulse from the Earth. This pulse comes from the solar system reaching here from the galaxy, which ultimately comes from the entire universe. We literally share this pulse with all of existence. Scientists have been recording the Earth's pulse for a long time, it has remained at approximately 7.8 cycles per second. This value remained constant until the year 1986-1987, when it began to increase until registering around 9 cycles per second in 1996. In a decade it increased by 1.2 cycles per second. In 2012, this pulse is expected to reach 13 cycles per second, as indicated by the Fibonacci theory.

Just as cymatics has shown that increased frequencies give rise to more complex patterns, we are now experiencing the beginning of a major shift in physical and spiritual vibration. « *It is difficult to understand what exactly will happen to our physical body, but ancient writings of pagan religions, monotheists, mystical groups and secret fraternal orders, offer some indications of what this experience could be like.* » (Esoteric Agenda) A military experiment that consisted of taking a bit of human DNA – a tiny piece of tissue taken from the mouth of a volunteer patient – and placing it in a device, which can measure the reactions of said DNA, which in turn I would be in the room of a building. The donor of the tissue from which the DNA was taken was in another room belonging to the same building.

Therefore, the living DNA of a person was in one room and the volunteer was in another room. They put the patient under a kind of emotional stimulation to receive from him a specific response: sadness, pain, joy, anger, etc. With this they wanted to see if the DNA from the other room had, in turn, an answer. Nothing in Western science would have supposed that this would work. But what they realized was exactly the opposite. All the emotions that the patient had in his room simultaneously reacted in the same way on the DNA that was isolated in the other room. The DNA response was expected to be later. However, there was no time lapse despite the distance, rather it was immediate. They ran the experiment again, but this time the donor was 400 miles from the DNA sample, and the result was the same. The donor was in Los Angeles and the DNA was in Phoenix, Arizona.

Another investigation consisted of studying the possibility that the heart had an electromagnetic field around it. This electromagnetic field, which protrudes between 5 or 8 feet outwards, would have the function of spreading health around the body. So they carried out the experiment, again through emotions, good or bad, sad or happy, hate or love, also through a DNA sample. They got the DNA particle to relax completely through the person's emotion of peace, love and relaxation. What we know from other experiments is that this relaxed state in DNA even responds to our immune system. Another procedure of the experiment led them to have the patient express anger, resentment, hatred, etc. which caused the DNA particle to shrink, the opposite of the previous experiment when it had been elongated (relaxed). Even in this experiment, some phases had been turned off by shutting down the immune system. The conclusion reached is that our state of mind influences the shape and response of our DNA.

MODERN BIOLOGY

One of the world-renowned and prestigious experts in the study of the origin of life: William Ford Doolittle, writes in his article ("New Tree of Life", Research and Science, 2000), the following conclusion:

> ◈ The reasonable explanation of such contradictory results must be found in the process of evolution, which is

neither linear nor so similar to the dendriform theoretical structure that Darwin imagined.

◇ The Darwinian view of the gradual evolution of organisms through random change does not work for the single-celled organisms that are the origin of life.

◇ The origin of eukaryotic cells (which constitute animal and plant organisms) occurred through the aggregation of different types of bacteria that currently constitute the nucleus and cell organelles, whose highly conserved gene sequences can currently be identified in the 4 kingdoms of life: protists, animals, fungi and plants.

This fact alone destroys the vision of the evolution of life as a phenomenon of gradual change, in which "random mutations" would be fixed or eliminated by natural selection -explaining it better:

1. This far-reaching change was not gradual.
2. If the mutations were random, the DNA of our cells would have very little to do with the bacterial one after more than a billion years of evolution.

William Ford Doolittle ends his article by refuting the Darwinian idea of a "tree of life" with a single ancestor, and concludes by saying that *"the data show that this model is too simple. Now new hypotheses are needed whose ultimate implications we have not even glimpsed"*, adding that *"the victory of Darwinism has been so complete that it is a shock to realize how empty the Darwinian view of life really is."* In another area, called Symbiogenetic Theory, its creator, Lynn Margulis –a leading internationally recognized American biologist- also makes Darwinism pale. Margulis's symbiogenetic proposal clashed (and still clashes today, although it has already been accepted as a specific fact) with the neo-Darwinists: The neo-Darwinian thesis that the evolution of organisms and the appearance of new species has its origin in errors in DNA replication (random mutations), falls before the results of this new research.

For Lynn Margulis, Darwinian natural selection continues without answering the source of evolutionary novelty, defending symbiogenesis: «*For more than forty years I have repeatedly heard of genetic errors. Genetic errors exist, but no species is known to have arisen through genetic errors. However, I observe numerous cases of symbiogenesis.*» What Margulis postulates is another new theory, which, although it once again exposes Darwinism, still cannot show the pure nature of each animal species respecting their typologies and family orders. What these professors do reflect is that genetic alteration or evolutionary changes to DNA are unlikely and, in fact, false.

Part IV

SCIENCE AND FANTASY

"We are here to learn to love each other.
I don't know why the others are here."
(WH Auden)

REGRESSIVE EVOLUTION

If we were to make the evolution hypothesis viable, another contradiction would cross our path. There are animals that represent a Regressive Evolution, that is, instead of improving, advancing towards a more developed species, they lose faculties, as if they were going backwards in the process of change. Some examples of this are:

1. From Mammoth to Elephant.
2. From Gigantopithecus to Gorilla and from Gorilla to Man
3. Of the Smilodon (Sabertooth)

The mammoth was woolly, huge, and had very long tusks. However, a process was changing it, as well as its cousins like the mastodon. Suddenly, these animals appear, losing superiority in their development, losing height, size in their tusks and fur. Something

183

similar occurs with man, if we follow Darwin's imaginary line, where giant apes reduce their size to that of a gorilla, and the latter to a human being, which would be a similar example of megantropus to man. Then, the other case, that of the tiger, which lost the elongated size of its "saber" fangs. This process is commonly called "involution."

Parallel Evolution

Because the vast majority of animals have not undergone any significant change, the theory of parallel evolution falls under its own weight. We have the same prehistoric animals today, except for the large reptiles that in the present would be a danger to our coexistence, and we still see prehistoric animals identical to those that survive today, within the order of their families and typologies. Obviously, just like us, all these living beings have other types, races, sizes and features, but they belong to the same animal families, those of the present with those of the past. Summarizing, in no case are they ancestors of each other, but rather they would be "cousins".

Suppose, for example, that 350 million years ago a fish tried to get out of the water. Evidently he would have died within a few seconds. Let's say that another fish, after the first, would have done the same. What would have happened? Obviously, it would obtain the same result as the previous one and it would continue to happen successively if more fish tried it, since it is known that fish have gills prepared only to carry out the specific respiratory process for the environment in which they live (remembering that we are talking about fish and not marine mammals). In the inconceivable event that a fish were magically transformed into an amphibian by a miraculous process of millions of millions of tries over millions of millions of years, the fish's chance would have changed nothing about other fish in other parts of the world. If that fish turned into a frog - for example - how is it possible that there were also frogs in the rest of the globe?

A surprisingly random process could have created another creature, but why the same toad? Did any toad or fish tell other fish on the other side of the Pacific and Atlantic so they knew what the "trick" was to become toads? By this rule, the same animals and the same families could not exist in different parts of the planet. It can be justified that there is the same variety of fauna and even marine flora, and likewise of flying birds, but it cannot be justified that there are the same amphibians, much less the same mammals and reptiles.

BERING'S EXODUS

The Bering Strait hypothesis is just as ridiculous as most of the ideologies postulated to try to justify the appearance of animals at the ends of the world. Accepting for a moment the ideology of parallel evolution both in America and in the Old World, we still could not say that plants, mammals and reptiles - mainly - crossed this strait from Siberia to Alaska, not even after an ice age. That is too much of a coincidence and a fleeting probability. Who led them down that route? Why would they risk moving forward with such a risky and unknown company? Is there any evidence to support the philosophy that animals passed through when the Bering Strait was united? And even if it were due to the action of ice in very cold times, does anyone know what kind of reptiles live in frozen lands? How did most dinosaurs get through the Bering Strait, since reptiles are cold-blooded and wouldn't survive a freeze? It's like taking a fish out of water and hoping it survives.

On the other hand, and taking reptiles as an example, they simply could not survive such a stage as the Ice Age did because they are cold-blooded animals. It is true that they have the advantage, with respect to warm-blooded animals, the fact that they need less food consumption to generate energy and that, in turn, last longer without eating. But they need to regulate their temperature, which they do with the ambient temperature. They need a climate in which, despite the possible cold at night, at least during the day there is a sufficiently high climate to be able to "charge their body batteries". We are not talking about a difficult climate of low temperatures without more, let's be real. We are talking about the Ice Age. Even thinking about the possibility that the reptiles had to endure a period of time in which they have to go through a cold zone on a constant basis is difficult to overcome. But going back to the facts and being realistic, a cold zone does not compare to what it could have been said era.

So it would be necessary to consider not only how long they have been "migrating" to get to where they had to go and, therefore, the average life of each species of reptile, but also the fact that in conditions like this the animals simply do not mate – Let's remember that they are in full emigration. The Ice Age simply could not have been feasible for the survival of this category of the animal kingdom. As a last and small detail, add that during that period the animals that inhabited the Earth were mostly reptiles. How did the animals get to Hawaii, Australia, Madagascar, Japan, and the rest of the world's islands? How did the supposed archaic man arrive? These and more questions are the ones that people do not usually ask themselves before consenting to the absurd accepted paradigm.

***

GEOLOGICAL COLUMN IMPOSSIBILITY

The theory that he believes he can describe a process by which all forms of life evolve (the word "evolution" means: "unfolding") from organisms that lived millions of years ago, invented by Charles Darwin (1809 -1882) and Gregorio Mendel (1822-1884), cannot seriously explain or justify the existence of most animals both in the past and in the present. They themselves considered that a period of chemical evolution precedes that of living organisms, since these "chemical" changes can only be carried out through "genetic manipulation" or through an "intelligent mind". Continuing to believe that we come from the monkey is an argument that has neither head nor tail. We are considered by conventional science as mammals and, of course, we fall under the name of "first mammals" (Pri-mates), along with tupaids, lemurs, lorisids, tarsids, monkeys and anthropoids, which, according to evolutionism, this group appeared 60 million years ago in the Paleocene, from insectivorous creatures.

According to this speculation, the prosimians derived from these insectivores and, in the Eocene -50 million years ago- the anthropoids, including the genus "homo". This is simply ridiculous, given that evidence of human life has been found tens of thousands of years ago, as we will cite later. What does the log actually show? Each new form of plant or animal—fern, shrub, tree, fish, reptile, insect, bird, or mammal—appears suddenly in the geological column. Immediately above the lifeless sediments of the Azoic Age, the Cambrian layer shows an abundance of crustacean and mollusc fossils, in large numbers, already fully developed; woody-stemmed plants appear suddenly in the middle of the Paleozoic Era. Fossilized wood has not been found in the lower strata, but it is abundant in later ages. Large collections of insect fossils have been found in upper Palaeozoic rocks, fully developed insects and in great variety, but none have been found in earlier strata.

At the beginning of the Cenozoic Era, mammals of modern types suddenly appear; there is no record of them having evolved from earlier types. Much importance is attached to the fact of finding "transient" evidence between one species and another to justify their evolution. But let's take the lungfish as an example, which has gills to breathe underwater and also a membranous bag that acts as a lung when it rises to the surface. At first you might think that we are facing one of those highly sought-after transitions.

But we find ourselves with the fact that this lungfish, which was supposedly in a process of evolution, never became a reptile or any other more "advanced" species. It is the same fish that still exists today and it is the same fish that is recorded in ancient fossil records. Definitely and based on the evidence, we see that it is not about any stage in evolution. It is simply a species created separately and has not gone extinct.

The evolutionary process is described as "the constant change of living things". However, many fossils have been discovered in ancient strata that, like the lungfish, are completely the same as the modern species we know today. The evidences or impressions left by the leaves of oak, walnut, American walnut, vine, magnolia, palm, etc. over time and found in rocks from Mesozoic and later times, do not differ from the leaves of these same trees and shrubs today. Geologists say that since these trees first appeared, estimated to be millions of years, there has been no evolutionary change in any of these species. Moreover, it should be noted that just as the leaves left evidence of these Mesozoic rocks, so did hundreds of insects, of which no evolution has been found.

These impressions show that these insects were very similar to the species of the same insects that exist today. As the evolutionist declares: " *The evolution of insects was essentially complete by the end of the Mesozoic* "...the era in which they originally appeared. Evolutionists now admit that the fossil record does not support the theories they have long espoused: " *There is no pattern that for the past 120 years we were told to look for,*" a paleontologist told a conference of evolutionists in Chicago, USA, in 1980. The picture of small changes accumulating to form new species is false. Rather: " *For millions of years species remain unchanged in the fossil record, and then they abruptly disappear, to be replaced by something that is considerably different, but clearly related,*" said a certain Harvard University geology *professor*. Individual species in the fossil record are characterized by stability, not change.

In order to have all the points on the table, we must know that the supposed pre-human skeletons have been manipulated and so have the rest of the discoveries to fit the Accepted Paradigm. The geological column, for example, is an international system developed in the 19th century to establish the age of fossils and of the Earth's strata, based on the idea that organic evolution was an established fact. However, the original classification of the strata was entirely arbitrary, as the fossils they considered to be the simplest were placed at the base of the spine, while in the upper strata they placed the fossils that supposedly implied more complex life forms. Against the geological column there is an impressive variety of evidence, among them we can see:

1. The geological column theory is a clear example of circular reasoning. The only reason for placing the rock formations in chronological order is the assumption of evolutionary progress; the only justification for assigning fossils to specific periods in that chronological order is the supposed evolutionary progression of life. To find out the age of a fossil, the scientist sees what stratum it is in; To find out the age of the stratum, the scientist looks at the types of fossils it contains.

2. Today it is known that even marine fossils that ancient geologists considered to be the simplest forms of life - trilobites, for example - are actually as complex as the organisms we see today.

3. It is important to know that the geological column does not exist anywhere in the world. A place with the geological column of fossils has never been found. It exists only in the minds of evolutionary geologists. It is simply a

representation, an ideal series of geological systems, and not a column of rocks that can be observed in a specific place.

4. Evolutionary geologists do not know how to explain the finding of fossils in the wrong stratigraphic order, or polystrat fossils. A typical example is the trunks of large trees that cut through several layers (National Geographic, p. 245, August 1975).

5. Scientists cannot explain the interruption in the deposition of sedimentary sequences or unconformities, which occurs whenever fossils are found in alternate layers, with gaps or gaps in intermediate layers.

6. If the geological column were true, men and dinosaurs would never have lived together. However, as we will explain later, cave paintings of men with dinosaurs, prehistoric human utensils and skeletons of men from the age of the dinosaurs have been found.

7. According to the geological column, trilobites constitute one of the first multicellular organisms, distant from man for millions of years. However, fossils of trilobites crushed by human sandals have been found. In reality, the evidence is contradictory for evolutionists.

Tree of Life

Let us remember that Darwin's tree of life is one of his pillars to explain evolution and its mechanisms, natural selection through mutations and other processes. British naturalist Charles Darwin's "tree of life", which shows how species are interrelated throughout the history of evolution, is flawed and should be replaced by a better symbol, according to a biologist at France's top science complex." *We have no proof that the tree of life is a reality,*" says Eric Bapteste, an evolutionary biologist at the Pierre and Marie Curie University in Paris, in statements to the magazine "New Scientist." Darwin devised in 1837 an imaginary tree to show how species could have evolved, a tree that quickly came to symbolize the theory of evolution through natural selection. Modern genetics, however, has shown that depicting the history of evolution in the form of a tree can be misleading, and many scientists argue that it would be more realistic to use a kind of impenetrable copse (a place populated by trees and shrubs) to represent the interrelationships. between species.

The genetic tests carried out on bacteria, plants and animals reveal that the species are interrelated with each other much more than previously thought, with which the genes are not only passed to the offspring through the branches of the tree of life, but are also transferred also from one species to another. Microbes exchange genetic material so promiscuously that it is difficult to distinguish one type from another, but plants and animals also interbreed very regularly, and the resulting hybrids can be fertile. According to some estimates, 10% of animals regularly create hybrids by crossing with other species. Do the fossils show the progression from single-celled structures, such as amoeba, to complex organisms? Let us consider the following facts:

1. Non-cyclical appearance of animals: The different and basic types of animals appear without responding to a chronology in the strata, without proof of ancestors. *«Evolution requires intermediate forms between species and paleontology does not provide them.»* (David Kitts, paleontologist and evolutionist)

2. Unaltered animals: In the evolution of animal "X", fossils of its evolutionary steps should have appeared: "X1, X2, X3..." If evolution is true, why doesn't that happen? Evolutionary history is supposed to have been filled with time-varying biological remains, as implied by the hypothetical aeonic transit from "amoeba to man." Contrary to common belief,

most fossils do not correspond to extinct animals, but are very similar (and often completely identical) to creatures existing today. There are many more living species of animals whose types are only known through their fossils.

QUESTION THESIS ON *the Extinction of Dinosaurs*

We have another question about those beginnings, and that is that, if evolution really occurred, then it would be rare for "extinctions" to occur. All beings on the planet would adapt and we would strangely know that term. Technically hundreds of billions of years ago there were no human beings to affect the characteristics of this planet so much. Even so, an extinction implies that a small number of creatures remain, of which a few others would be the ones that "causally" would have evolved, and that makes the evolutionary hypothesis less feasible. There were several cataclysmic events in the past, so how do you know for sure that the great Triassic extinction wiped out 96% of marine species and 70% of land species? If one speaks with such certainty, it means that they know all that existed and also that it has been possible to keep track of all those that have continued to evolve until now, to differentiate them from those that have disappeared, but this is not the case.

«*In science, as in many other areas of knowledge, it is very important to make it clear what we mean by each word we use. When we talk about evolution we should clarify what kind of evolution we mean, since this hypothesis tries to cling to many different parameters in its quest to justify itself.*" (Abiam F. Palomares, researcher and historian). The extinction of the dinosaurs 65 million years ago cannot be explained solely by an asteroid colliding with Earth, but is the result of a long process of climate transformation, according to research results revealed by a German paleontologist. The asteroid was just " *the latest catastrophic event* " following " *at least 500,000 years of massive climate fluctuations"* that severely weakened the ecosystem, said paleontologist Michael Prauss of the University of Berlin. The American scientific magazine "Science" reported the work of a group of scientists who attributed the disappearance of the dinosaurs to a gigantic asteroid that fell in the current Mexican region of Yucatan.

«*Contrary to the publication in Science, which did nothing more than bring together already known elements, my work is based on new data [...] that allow us to reconsider everything from a new point of view*», Prauss said. The German paleontologist has been working with an international scientific team since 2005 as part of a project of the German Agency for Scientific Research (DFG). This team analyzed rocks and samples from a 25-m-deep borehole in Texas, USA, 1,000 km northwest of the asteroid crater. The work made it possible to demonstrate the existence, long before the impact of the asteroid, of important climatic transformations, *"probably caused by volcanic activity"* that took place over several million years in present-day India, says a statement from the Free University of Berlin. According to Prauss, *"the long-term climatic stress produced by this, to which the meteorite strike evidently contributed in the end, explains the crisis of the biosphere and the mass extinction"* of species in the Tertiary Cretaceous. (AFP, Berlin. March 19, 2010. www.prensalibre.com/vida/ciencia/Cuestionan-tesis-extincion-dinosaurios_0_227977339.html [1])

⁕

IMPOSSIBLE FABLES

It is absurd that evolution exists if there is extinction for other important reasons: if a process of evolution led an organism to become another immediately superior over millions and millions of

---

1. http://www.prensalibre.com/vida/ciencia/Cuestionan-tesis-extincion-dinosaurios_0_227977339.html

years, it should have been without stopping. In the case of man, if he came from the monkey, the monkey from the shrew, the shrew from a type of dinosaur, and this from an amphibian in a period of 300 million years, we must add the fact that several mass extinctions occurred that they eliminated most of these creatures and the "chance" that they would evolve. The greatest extinction that occurred tens of millions of years ago wiped out nearly 95% of animal species, leaving little creature survival. These would be a minimum probability of the evolutionary fact, with little possibility of survival and adaptation.

Other subsequent mass extinctions wiped out over 65% of living creatures. So this is the same as starting from "0" from a few organisms of different categories, reducing the few species that remained in line of disappearance. The strongest would not be the ones that would mutate into something better but the ones that would simply survive the atmospheric conditions. That is to say, by the evolutionary rule of millions of years of adaptation and change, said event would not have given the species time for self-improvement and modification before their disappearance. The theory now would be how so many animals survived and adapted in "such a short time." We are talking about sudden evolution since there have been several ice ages and interglacials that eradicated life from the globe. How did it appear again? These species had to succumb and if any survived they had to do so in extreme environmental conditions for which they are not prepared and so fast that they would not have time to adjust.

The Last Ice Age, which occurred about 12,500 years ago, started our world anew: plants, reptiles, amphibians, birds, mammals, and other organisms. How could they reappear and evolve in a matter of years? The fossil record makes it clear that there have always been animals, so someone must have brought them here or in a "creationist" way, made them. But reptiles, mainly, would not have survived a frost. «*What is it that has exterminated so many species and entire genera? The mind is at first quick to believe that there was some great catastrophe, but to destroy animals, both large and small...we must shake the whole framework on the globe. No less physical event could have brought about this total destruction, not only in the Americas, but throughout the world... Indeed, no event in the long history of the world is as astonishing as the widespread and repeated extermination of its inhabitants.*" (Charles Darwin, quote from his diary, Voyage of the HVS Beagle)

THE HUNDREDTH MONKEY *Theory*

About the relationship between living beings, someone said: «*If there is no evolution, living beings are not related to each other, since they do not have an evolutionary relationship, if they do not have an evolutionary relationship, then they cannot have any common protein, since that being different entities, they should not share any character. According to whose criteria or what are these conclusions drawn? Does the fact that a living being shares cellular similarities with another already have an evolutionary character or is it related to that species? Who says it? We live in a similar global ecosystem and adaptation to this planet requires similarities in species and even more so at the cellular level. We had to adapt to this planet from the beginning, that is, we already had the necessary conditions to survive on this globe from the beginning. The truth is that the conjectures about the morphologies in the species in reference to their possible similarities continue to be studied because to this day it is not demonstrable that by sharing cellular similarities they are ancestors of a species.*» (Abiam F. Palomares. Researcher, writer and historian)

If there is one thing that evolutionary theorists cannot explain, it is the fact that animals share a close spiritual bond no matter how far apart they are. This invisible bond unites even humans, that is, it is shared by all living beings, potentially among the same races. The Hundredth Monkey Theory is about a team of scientists in the 1970s who worked on a particular type of island off Japan with a particular type of monkey. These monkeys were all over these island chains, and they were all being monitored. They ate a type of tubers that filled with sand when they fell to the ground, and a monkey washed some of them and saw that they tasted better that way, and many of them began to tell each other about it and they all began to do so. When they reached this critical mass without anyone teaching them this they were already doing it, but this began to happen simultaneously on other islands where the monkeys could not communicate in any way.

Something like this was the first time in history that it had been seen, to the point that a book was written about it. Australian scientists, based on this, also made a study, but starting from the human race. They came up with this with a painting that would have many people painted on it. This worked, but the aborigines of Australia already knew it for a long time, they were aware that there is a structure between men that unites them. This occurred in many times with inventions and ideas, where someone claimed to discover or develop something new and elsewhere another person said the same thing having reached the same conclusion or revelation. The US and Russia discovered these electromagnetic nature fields 80 to 100 km above the surface of planet Earth.

In the world of mainstream science, and having come a long way from its initial use as an identifier for functional proteins, comparative genomics is now concentrating on finding regulatory regions and siRNA (small interfering RNA) molecules). It has recently been discovered that distantly related species often share long stretches of conserved DNA that do not appear to code for any protein. The functionality of such ultraconserved regions is unknown at this time. The spirit world is the sovereign reason why the world of Freemasonry has tried to deny God. Not knowing where we come from implies knowing what we are or who we are. We are considered "royalty" of "divine" blood under the concepts taught by the Jewish, Christian and Muslim religion, which forces us to behave as such: Children of God. Being creators like the ONE and responsible for the maintenance and recreation of the universe. On the contrary, believing that we are beasts and the product of chance exonerates us from any commitment to the spiritual "bond" that unites everything.

THE DINOSAURS OF YESTERDAY and Today

Who did the aquatic squirrel that lived with the dinosaurs millions of years ago evolve from, and what creature did it evolve from? Chinese scientists discovered fossil remains of a semi-aquatic mammal much earlier than thought for this event. This animal was already swimming and eating fish 164 million years ago! that is, with the first dinosaurs! Their resemblance to something we could identify today would associate them with our current beavers, otters, and platypus. Although it certainly wouldn't look much different from a huge rat with a platypus tail. Chinese scientists, who made the official discovery in 2004, recently presented it to the public in the journal Science. The researchers say the fossil skeleton shows that in the Jurassic - during the reign of the dinosaurs - some mammals occupied more varied ecological niches than previously suspected.

Thomas Martin, a specialist in early mammals consulted by The New York Times, said the new find pushes back "*the conquest of water by mammals by more than 100 million years,*" and contradicts the conventional view that the first mammals were insectivorous creatures. nocturnal creatures that did not explore the world until the dinosaurs disappeared. Other discoveries show that sharks, spiders and scorpions already swarmed the Earth 300 million years ago, they have not evolved. They remain entirely spiders and sharks. This is the case of the fossil of an arachnid that lived in the Carboniferous Era. With the help of a microscope, silk manufacturing structures similar to those used today by spiders in their webs can be observed. This shows that silk thread formation techniques are older than previously believed. The finding belongs to the Aphantomartus pustulatus from 300 million years ago. (Source: Cary Easterday – OSU. Very Interesting - No. 283. December 2004).

A conceptual problem with the hypothesis of the evolution of species lies in the fact that practically all animals in the past remain the same as in the present. That is, they have not undergone any change or process. In this sense we can find prehistoric animals even in the present. Among some cases of these antediluvian creatures, their genera and families, we see:

· Algae: Stromatolites (have not changed in 3.3 billion years)

· Antelopes: Antilocapra.

· Ostriches: Diatrymas, Brontornítidos, Hesperornis, Phororhácidos and Odontognatas.

· Horses: Orohippus, Diadiaphorus, Hipparion and Eqqus.

· Camels: Macrauchenia and Alticamelus.

· Crabs: Notopocorystes.

· Ammonite-snails: Amalteus.

· Civets-mongooses.

· Crocodiles.

· Corals: Hippurites and Zaphrentido.

· Cockroaches.

· Dragons of Comodo: Varanosaurus.

· Elephants: Mammoth, Mastodon, Paleomastodon and Anancus.

· Trilobite scorpion: Eurypterid.

· Sea stars.

· Gulls: Ichtiornis.

· Peripatus worms: Aysheia; Worms: Onychophorans.

· Lizards: Hylonomus.

· Hyenas: Percrocuta.

· Hippos: Diprotodon and Toxodon.

· Ants

· Giraffes: Helladoterium and Okapis.

· Dragonflies: Meganeurae.

· Sea lilies: Crinoids.

· Manta ray: Aëtobatus and Jackdaw.

· Marsupials: Thylacosmilus and Opossum.

· Jellyfish: Dawsoni jellyfish.

· Molluscs: Cephalopods (they have not undergone any modification for 225 million years)

· Monkeys: Notarcus, Australopithecus Homo Habilis, "Ida", Plesiadapsid and Ramapithecus.

· bats

· Oysters: Cardium, Schizodus and Steblochondria.

· Dogs: Borhyaena and Thylacinus.

· Pumas and other cats: Smilodon and Machairodus.

· Rhinos: Diceros, Arsinoitherium and Uintaterium.

· Grasshoppers and crickets: Neoptera of the "Palaeoptera" group.

· Sardines: Synodus.

· Snakes: Serpentoid Lepospondylians.

· Sharks: Cladoselache, Galeocerda and Odontaspis.

· Turtles: Triassochelys, Proganochelis and Archelon.

What evolution is there in this, if the vast majority of these creatures are of the same line as today's animals, and their only difference is the same as a pony with a horse or a donkey with a zebra? The variety of creatures spread throughout the world excludes the possibility that they had a common relative: we see capuchins and spider monkeys in Central America; the howler monkey and the marmoset in South America; the Berber monkey in the Canary Islands (Spain); baboons, gorillas, chimpanzees, monkeys and colobus in Africa; the macaque in the Middle East; the gibbon in China and orangutans in Oceania. How were they distributed like this? There are very particular animals, such as flying ants, the flying dragon, and common species in rain forests such as flying snakes, giant flying squirrels, the flying flagellum, the flying gecko and the flying frog.

In each species, the creatures are represented in their multiple typologies, as if an intelligent designer had been inspired to decorate each one of each family with different colors and special characteristics. I will cite some significant and very recent examples of modern animals that appeared millions of years ago and have not changed their genetic or physical structure at all. Mostly these belong to a discovery made in Poland:

SPIDER

Age: 50 million years

Period: Eocene

Location: Poland

Countless fossils belonging to different species of spiders show that these arachnids have existed in their perfect form, with all the characteristics they now possess since the moment they began to exist. None is semi-developed. Of all these, not one has been discovered that has become another form of life. To put it another way, spiders have always existed as spiders, and will always exist as spiders. The clear example is a spider preserved in amber that is 50 million years old, and shows that, like other living things, spiders never evolved.

APHID

Age: 50 million years

Period: Eocene

Location: Poland

The aphid is a species of insect that feeds on plants and is a member of the superfamily Aphidoidea. There are about 4,000 known species of aphids, divided into ten families. The oldest aphids identified so far lived in the Carboniferous period (354 to 290 million years ago). They have not changed in the slightest in the more than 300 million years that have elapsed. The 50-million-year-old aphids preserved in amber prove that these insects haven't changed since the moment they were born—in other words, they haven't evolved.

FLY

Age: 50 million years

Period: Eocene

Location: Poland

One of the most distinctive features of the fossil record is how species remain the same during the geological periods in which they appear. A species preserves the structure it had when it first appears as a fossil until it either becomes extinct or reaches the present unchanged, over the course of tens or even hundreds of millions of years. This is clear proof that living things never evolved. There are no differences between the 50-million-year-old fly fossilized in amber found in Poland and the flies that live today.

PSEUDOSCORPION

Age: 45 million years

Period: Eocene

Location: Russia

These arachnids, belonging to the edge of arthropods, have received this name because their structure is similar to that of scorpions. However, their anatomical features are much more similar to spiders than scorpions. The oldest known specimens lived in the Devonian period (417 to 354 million years ago). These invertebrates have never changed since the time they first appeared in the fossil record.

other insects

Age: 125 million years

Period: Jurassic

Location: Liaoning Province, China

Contrary to what evolutionists claim, these insects, several species of which are found as fossils from the Carboniferous period (354 to 292 million years ago), have no evolutionary ancestors. Each species appears abruptly in the fossil record with its own structures and characteristics, and they remain the same throughout its existence. This fact makes it impossible for evolutionists to defend their view of evolution.

COCKROACH

Age: 125 million years

Period: Lower Cretaceous

Location: Liaoning Province, China

Cockroaches live anywhere on Earth, with the exception of the polar regions, and can be traced in the fossil record for millions of years, with their perfect, fully developed structures. Cockroaches, having preserved their structures from before 125 million years ago, announce that they never evolved, but were created as we know them.

FROM THE GUT TO THE Lung

Heribert Nilsson was a Swedish botanist aware that because there are now more than a hundred million fossils, all cataloged and identified in museums around the world, and after 40 years of research, the evolutionary transition cannot be believed. And is that, what happens with the respiratory system of amphibians evolved according to the hypothesis? At what point did a fish go from having your type of respiratory system to that of an amphibian or a reptile? There is no evidence to suggest that such a "miracle" of nature could have happened where a marine organism changed its entire internal apparatus to constitute a totally different one such as that of land animals. A fish filters the oxygen from the water through its gills and from there it goes directly into the blood, on the contrary, amphibians and reptiles literally breathe air through their nostrils and it passes to the lungs - an organ already fully developed - and from the lungs it passes into the blood.

Fish do not have lungs, with the exception of aquatic mammals, such as dolphins and whales. Does anyone know the same fish as beef, or chicken with pork chop? It has nothing to do with it. In addition to this, if the evolutionary scenario suggests that a life form advances to an immediately superior one. What need led the fish to want to live out of water? Didn't the ocean provide them with their necessary requirements? What force pushed them to live outside their habitat? And why evolve outside of your environment, when you can improve within your own environment? The chances of surviving in a hostile and unknown environment are less than those of staying in the natural ecosystem, that's a no-brainer. For fish, the outside world would be a probability of extinction rather than survival, adaptation and fruition.

FAKE EMBRYO DRAWINGS

The evolutionary biologist Ernest Haeckel at the end of the 19th century postulated that living embryos re-experienced the evolutionary process that their false ancestors followed. Haeckel theorized that during the development of these in the womb the human embryo first shows the characteristics of a fish, then those of a reptile and finally those of a human. Ernest Haeckel, published several drawings to support his idea. The truth is that Haeckel made false drawings to make it appear that fish and human embryos were alike. He claimed that human embryos have gill openings, which led to the belief that man had come from fish. However, those folds of skin are not gills , as they end up developing as bones in the ear and glands in the throat. Rudimentary organs such as the coccyx make us believe that we come from animals with a tail, however, there are 9 muscles attached to the coccyx, which are not rudimentary. When it was discovered, all he could say in his defense was that other evolutionists had done similar things.

After the compromising confession of these "falsifications" anyone would want to condemn and annihilate him, if it were not for the consolation of seeing hundreds of other culprits at his side, including trusted observers and other more renowned biologists. . (See Creation Seminar Video #4. Also New Scientist, Sept. 6, 1999. Pg. 23.) Most of the diagrams that appear in the best biology books, treatises, and magazines are guilty of falsification to the same degree, since they are all inaccurate and to a greater or lesser extent manipulated, schematized, and retouched. It is interesting to note that although Haeckel's falsification occurred in 1901, in many evolutionary publications the subject appeared for a century as if it were a proven scientific law. Another example is that of whales, where it is believed that they have a rudimentary pelvis, which suggests that they came from a terrestrial creature. Ergo, those bones are support points for the muscles. Without them, the whales could not reproduce. If there were "rudimentary organs" that is "losing something", that is, it is the opposite of evolution.

THERE ARE NO ANIMALS *as Transitional Forms*

Evolutionists imagine that there was a type of animal that slowly changed over a long period of time (millions of years) to become a different type of animal. For example, they believe that amphibians gradually became reptiles through this process. If we think about it, this would indicate that millions of creatures must have existed during this time that would be "in between", while amphibians evolved to become reptiles. If the theory of evolution has any truth to it, then the evidence for these "transitional" forms should be abundant. There must be many fossils that are part amphibian and part reptile. The surprise is that many experts in the study of fossils acknowledge that no transitional form has been found anywhere in the world among any group of creatures. Have museums ever seen fossils of creatures that are 10% dinosaur, 90% something else? 20, 30, 40, 50%? 90%?

The fossils found are 100% dinosaur, there is no evidence of transitional forms. Among all the fossils discovered over the years, there is not a single example of the intermediate forms that would be necessary if, as the evolutionary hypothesis defends, living beings had evolved step by step from simple species to more complex ones. We see that Robert Carroll, expert in vertebrate paleontology and convinced evolutionist, has admitted: «*Despite more than a century of intensive data-collection efforts since Darwin's death, the fossil record still does not offer the picture of infinite links of transition he expected.*" The major groups of animals appeared at once and fully formed in a very short period of time known as the Cambrian Exposure.

Before that time there are no remains in the fossil record of another organism but primitive unicellular and some multicellular creatures. All the fossils found in the Cambrian rocks belong to very different creatures, such as snails, trilobites, sponges, jellyfish, starfish, mollusks, etc. Many of the creatures in this layer have complex systems and advanced structures such as eyes, gills, and circulatory systems, just like modern specimens. These structures are very advanced and at the same time very different. The process that would transform molecular beings into multicellular beings, invertebrates into vertebrates, fish into amphibians, amphibians into reptiles, reptiles into birds, oviparous mammals, etc., does not occur in any type of species. That is to say, scientifically, practically all evolutionary traits of this type have been denied in the beings of our globe.

Citing an example, let's talk a little about the Coelacanth. Based on fossil evidence, evolutionists believed that it was an intermediate between fish and amphibians. The reconstructions showed the Coelacanth with amphibious and ictine characteristics. Subsequently, living Coelacanths were discovered in the Indian Ocean near the Cape Province, in southern Africa. They were fish. The reconstructions had been wrong. And it is that, if one is only based on conjectures, nature shows one his gross mistake.

ADAPTIVE CHANGES IN Birds?

A pamphlet published in 1999 by the US National Academy of Sciences describes Darwin's finches as "a particularly compelling example" of the origin of species. The booklet cites Gran's work and explains how *"a single year of drought on islands can lead to evolutionary changes in finches."* The booklet also calculates that " *if there are droughts about every 10 years on the islands, a new species of finch could emerge in a mere 200 years* ." But this booklet silences the fact that the finches' beaks reverted to normal after the rains returned. There was no net evolution. In fact, there are several species of finches that currently appear to be mixing through hybridization, rather than diverging by natural selection as Darwin's theory demands. Suppressing evidence to give the impression that Darwin's finches confirm evolutionary theory borders on scientific malpractice.

According to the Harvard biologist, Louis Guenin (Nature magazine, 1999), if plants in consensus with light are already capable of showing us that they use states of superposition to choose how to make the best use of their energy pathways, any living being will be able to do so even more. without the mediation of chance or the imposition of the environment; that is what is perfectly extrapolable. What a process as relatively simple as photosynthetic and superposition, but so crucial, and that it is not even animal, is telling us is that any state of superposition is capable of choosing. Then responds to approach and choice.

EVOLUTIONISTS VS EVOLUTIONISTS

A well-known British paleontologist, Derek V. Ager, although he is an evolutionist, admits the following: " *What emerges, if we look closely at the fossil record, whether at the order or species level, is that what we find time and time again is not It is not a gradual evolution but the sudden explosion or emergence of one group at the expense of another.*" (Derek V. Ager, "The Nature of the Fossil Record", Proceedings of the British Geological Association, vol. 87, 1976, p. 133). Another evolutionary paleontologist, Mark Czarnecki, that « *Fossil records, the traces of extinct species preserved in the geological formations of the Earth, have been a great problem for the demonstration of the theory. Such records have never revealed traces of Darwin's hypothetical intermediate variants. Rather, species appear and disappear abruptly, and this anomaly has encouraged creationist arguments that each species was created by God.*" (Mark Czarnecki, "The Revival of the Creationist Crusade," MacLean's, January 19, 1981, p. 56).

They have also dealt with the futility of the "lost" transitional forms turning up in the future, as a University of Glasgow professor of paleontology, T. Neville George, explains: "There is no need to apologize for any longer than usual. *"poverty of the fossil record. In a way they have become almost unmanageable because of their sheer size and the discoveries are putting integration out of the question...*

Yet the fossil record continues to be made up mostly of gaps." (Neville George, "Fossils in Evolutionary Perspective", Science Progress, vol. 48, January 1960, pp. 1, 3). Adding more irregularities of said postulate, we highlight that the theory of evolution is against the first and second laws of thermodynamics. They say that *"everything has a degenerative process... "*, and evolutionary theory says that *" the natural process that took millions of years changed simple and disordered molecules into complex and highly ordered living structures "*. This is completely unscientific. This law of thermodynamics says that *" all spontaneous processes change from complex to disorder, and organized energy to heat energy."*

According to the general theory of evolution, the basic progression from life culminating in man was: inert matter, to protozoa, to metazoan invertebrates, to vertebrate fish, to amphibians, reptiles, birds, fur-bearing quadrupeds, apes, and finally the man If the theory of evolution were accurate we would expect to find a vast number of objectively preserved forms in the fossil record. Transitional forms are totally absent from the fossil record. Archeopteryx was once thought to be a transitional form, but has since been recognized by paleontologists as a true bird. Evolutionists, knowing this error in their belief system, now argue that no fossils are present because they were brief "evolutionary bursts" over billions of years, and because of their brevity and rapidity they left no trace in time. However, the belief in "evolutionary explosions" is still not supported by either the First or Second Law of Thermodynamics, or the Law of Bio-Genesis.

THERE IS NO TRANSIENT *Body between Invertebrate Beings and Fish*

Evolutionists accept that marine invertebrates on record in the Cambrian stratum somehow evolved to become fish over millions of years. However, precisely because Cambrian invertebrates do not have any ancestors, there is no link that justifies such a transformation. It must be added that invertebrates and fish have many and abysmal structural differences. While invertebrates have hard tissues on the outside, vertebrates have those tissues on the inside. If this evolution had been a fact, endless fossils and evidence of each of the transitions would have been found throughout history, of which not one has been found. Evolutionists have been excavating the fossil strata for nearly 140 years after finding these hypothetical forms.

They found millions of fossil invertebrates and millions of fossil fish. However, no one has found, even if it is not, a fossil halfway between the invertebrate and the fish. At what point did an animal with an exoskeleton become an endoskeleton? An evolutionary paleontologist, Gerald T. Todd, admits this fact in an article entitled: "The Evolution of the Lung and the Origin of Bony Fishes": «All three *subdivisions of bony fishes appear for the first time in the fossil record more or less at the same time. They already appear morphologically very differentiated and are well armored. How did they originate? How did they come to have a strong shell? And why are there no traces of primary, intermediate forms?* » (Gerald T. Todd, "Evolution of the Lung and the Origin of Bony Fishes: A Casual Relationship", American Zoologist, vol 26, No. 4, 1980, p. 757).

THERE IS NO TRANSIENT *Creature between Reptiles and Birds*

As usual, there is no evidence of transitional forms that were supposed to link even amphibians with reptiles. Evolutionary paleontologist and authority on vertebrate paleontology Robert L. Carroll has to accept that " *early reptiles were distinct from amphibians and their ancestors could not yet be found*." (Robert L.

Carroll, Vertebrate Paleontology and Evolution, New York: WH Freeman and Co., 1988, p. 198). The theory of evolution maintains that the ancestors of birds were dinosaurs, members of the reptile family. Evolutionists have claimed that a bird called Archeopteryx represented this transition. However, the latest studies of Archeopteryx fossils reveal that this explanation lacks scientific foundation. Not a transitional form at all, but an extinct species of bird with insignificant differences from modern birds.

As the eminent paleontologist Carl O. Dunbar explains, " *by its plumage, it must be clearly classified as a bird.*" Despite the hopelessly failed scenarios, evolutionists are not yet done with their drawbacks. Since they believe that the birds must have evolved in some way, they claimed that the transformation occurred from the reptiles. However, none of the various mechanisms of birds, which have a completely different structure from terrestrial animals - including the bird Archeopteryx, whose peculiarity is also explained in Harun Yahya's book - can be explained by means of gradual evolution. First of all, the wings, which are the exceptional feature in birds, present a great difficulty for evolutionists.

One of the Turkish evolutionists, Engin Korur, confesses the impossibility of wing evolution: « *The common feature of eyes and wings is that they (only) can function if they are fully developed. In other words, a half-developed eye cannot see, a bird with a half-formed wing cannot fly. How these organs came into existence has remained one of nature's mysteries, a mystery that needs to be unraveled.*" (Engin Korur, "Gizlerin ve Kanatlarin Sirri" - The Mystery of the Eyes and the Wings -, Bilim ve Teknik, No. 203, October 1984, p. 25). It remains totally unanswered how the perfect wing structure came into being through consecutive random mutations. There is no way to explain how the front arms of reptiles could become perfectly functioning wings as a result of a distortion in the genes (mutation).

In addition, it is not enough to have wings for a terrestrial organism to fly, since many other structural mechanisms are needed that birds use for that purpose. For example, the bones of birds are much lighter than the bones of land animals. Your lungs work very differently. The muscle and skeletal systems are different, and the blood circulation system is tremendously specialized. These traits are prerequisites needed for flight, at least as much as wings. All these mechanisms had to be present together and simultaneously. They could not be formed gradually by "accumulation". For this reason, the theory that states that terrestrial organisms evolved to become aerial organisms is completely false.

SUDDEN EVOLUTION THEORY

Why have transitional fossils not been found? Evolutionist Jeffery Schwartz argued, " *these have not been found because they do not exist.*" ("Pitt. Professor's Theory...", 2006). Has this impartial evolutionist jumped ship and joined the creationist crowd? Absolutely not! In fact, his statement is simply a means to support an alternative evolutionary theory in order to argue the weakness of evolution. Schwartz argues for a new theory, titled "sudden origins," instead of the gradual, incremental changes once proposed by evolutionists. Schwartz argues that gradual change does not occur, stating that "evolution is not necessarily gradual but often a sudden and dramatic expression of change" ("Pitt Professor's...", emp. added).

Schwartz wrote an article in the January 30, 2006 issue of New Anatomist magazine. The University of Pittsburgh news release indicated that his paper gives a better understanding of the structure of the cell, which Schwartz argues provides solid support for his "sudden origins" theory of evolution. This "newly improved" or "cleverly packaged" version of evolution was originally detailed in Schwartz's 2000 book, Suden Origins: Fossils, Genes, and the

Emergence of Species. the species). According to Schwartz, evolution is an expression of " *change that began at the cellular level because of radical environmental stresses—such as extreme heat, cold, or crowding—years ago* " ("Pitt Professor's..."). The mechanism, Schwartz explains, is that " *Environmental turmoil causes genes to mutate, and those altered genes remain in a recessive state, spreading silently through the population until offspring appear with two copies of the new mutation and change." suddenly, appearing seemingly out of nowhere.* » ("Pitt Professor's,..."). Of nothing? Nothing creates nothing.

Defending his new model, Schwartz described why cells do not change subtly and constantly on a small scale over time—as Darwin and his followers predicted. The press release observed: " *Cell biologists know the answer: Cells don't like to change, and they don't do it easily."* Consequently, these massive environmental changes lead to mutations that "may be significant and beneficial (like teeth or limbs) or, more likely, may kill the organism" (emphasis added). Schwartz further argued that " *it's the environment that throws them off their balance and most certainly kills them eventually when it changes them. Therefore, they are being shaken by the environment, they are not adapting to it."* ("Pitt Professor's..."). Summarizing:

1. Transitional fossils, to show that one organism would have evolved into another, do not exist.
2. Gradual change doesn't happen—it's sudden. This supports the creationist theory.
3. Cells don't like to change and don't do it easily.
4. Mutations cannot provide a good "balance" to make the changes necessary for evolution to occur—and, even then, they will probably "kill" the organism.
5. Organisms are not adapting to the environment, but are reacting to it.

This sounds like a text written decades ago by creationists (who have long accepted that life appeared suddenly). Indeed, there are no transition fossils between one species and another and the gradual changes cannot be explained nor can they explain the diversity of life that surrounds us. How long will it take for these men to take the final step and give God credit for the "sudden origin" of life? A lot, because his interest is not scientific but atheistic. While these scientific studies continue to make headlines, is it any wonder that people are now questioning the theory of evolution? Many have come to realize that evolution does not deliver the answers that we all have and that were once, long ago promised.

In fact, BBC News reported that, "*according to a public opinion poll, only less than half of Britons accept the theory of evolution as the best description of the development of life.*" ("Britons Unconvinced...", 2006, emphasis added). In the BBC report, Andrew Cohen, editor of Horizon, noted: "*Most people would have expected the public to vote for the theory of evolution, but there seem to be a lot of people who apparently believe in an alternative theory for origins.*" of life ." The report went on to note: "*The discoveries caused the scientific community to be surprised. Don Martin Rees, President of the Royal Society, said: 'It is surprising that many are still skeptical of Darwinian evolution. Darwin proposed his theory nearly 150 years ago and it is now supported by a large body of evidence.*" ("Britons Unconvinced on Evolution," BBC News, January 26, 2006) Plenty of evidence?

Notoriously, Don Martin Rees needs to be aware of reality like Schwartz, and a host of other people. The actual "amount" sustains a sudden origin (which could only be explained by God's craft). There ARE no transitional fossils, no gradual changes, and no beneficial mutations. From the looks of it, it seems that the actual "great number" is the truth that evolutionists should accept.

MORE TROUBLES FOR DARWIN

In the complete absence of intermediate fossils, zoologist Harold Coffin says: "*If the concept of progressive evolution from the simple to the complex is correct, the Cambrian should find the ancestors of these fully developed living creatures; but they have not been found, and scientists admit that there is little chance that they ever will be found. On the basis of the facts alone, on the basis of what is actually found on Earth, the theory of a sudden act of creation in which the main forms of life were established fits the best.*" With no evolutionary ancestors, insects appear suddenly and in large numbers in the fossil record. In his book "On Growth and Form" Zoologist D'Arcy Thompson says: «*Darwinian evolution does not teach us how birds descended from reptiles, mammals from earlier quadrupeds, quadrupeds from fish, nor vertebrates from earlier the invertebrate branch... To search for stones through the gaps between them is to search in vain, forever.*»

Of all the missing links in the transformation of animals to more complex species, according to the theory of evolution, the link that would unite the ape with man is the most famous of all. « *The known fossil remains of man's ancestors would fit on a pool table. That makes a poor platform from which to try to penetrate the mist of the last few million years.*" Elwyn Simons, Duke University. « *The missing link between man and anthropoids [...] is simply the most attractive of a whole hierarchy of phantom creatures. In the fossil record, missing links are the rule.*" (Newsweek Magazine). Charles Darwin's son wrote: " *We have no record of any change from one species to another...we cannot prove that any species has changed.*" (Darwin, Francis, ed., The Life and Letters of Charles Darwin, vol. 1, p. 210) But this would not be the first nor the only one of Charles Darwin's headaches. And here are other famous phrases of the prophet of many atheists:

" *The distinctiveness characteristic of specific forms [of life], and the fact that they are not discernibly connected to one another by innumerable transitional links, is a very obvious difficulty."* (Charles Darwin)

« If numerous species have actually started their existence at once, that fact would be deadly for the theory of evolution. " (Charles Darwin)

«Why are not all geological formations and strata full of these intermediate links? Certainly geology does not reveal such a finely graded organic chain; and this, perhaps, is the most obvious and serious objection that can be raised against the theory." (Charles Darwin)

« The abrupt way in which entire groups of species suddenly appear in certain formations has been advanced by several paleontologists as a deadly objection to the belief in the transition of species. " (Charles Darwin)

« There is another difficulty, related to this one that is much more serious. I am referring to the way in which species belonging to several of the main divisions of the animal kingdom appear suddenly in the lowest known fossiliferous rocks." (Charles Darwin)

FANTASY OF THE PROCESS of Change in Animals

In the textbook entitled "Man and the Biological World", the authors admit that such proof is not complete: "The existence of homologous similarities, parallelisms in embryonic development and different degrees of chemical relatedness between organisms does not in itself prove that there has been evolution". In this way, we are asked to turn to the fossil record to find the final and decisive proof that evolution has really happened. One might imagine that we would find a series of fossils, for example, of mollusks whose exterior was initially hard and gradually turned into a scaly exterior, while part of the latter turned inward and became a spine (an exoskeleton). turned into an endoskeleton. In turn, successive fossils would show the development of a pair of eyes and a pair of gills at one end, and a tail shaped like what could be a fin at the other. Finally, here would be our fish!

But the fish would not remain fish. Rather, according to the geological column, we would be waiting for the fish to transform its fins into legs, from which feet and fingers would emerge, and its gills would become lungs. At a higher level, we would no longer find the fossilized remains of these organisms in ancient sea beds, but buried in dry land. On the other hand, other fish, the fins would become wings and clawed feet. The scales would become feathers and a horny beak would form around the mouth. It seems that in this way and after an "abracadabra" we would have, thanks to the magic of evolution, birds and reptiles. If things were like this, we would not stop finding endless transitional fossils of each of the existing species on the face of the Earth. But once again we come to the same conclusion and that is that a single piece of evidence for this "fabulous" theory was never found.

The links, aptly, have been called missing links. But the reality is that they are not lost, nor are they links, because they never existed. We also have the case of the African fauna of prehistory in the Iberian Peninsula. Science cannot explain the proliferation of innumerable North African fauna in the Iberian Peninsula, among them the "baboons", since archaeologists assure that there was no connection between the 2 shores of the Mediterranean for at least 5 million years. Who put them there?

Part V

MEN AND APES

*"Man is born free, and yet
everywhere is chained."*
(J.J. Rousseau)

THE FIRST MONKEYS OF the Novel

Henry Morris in his well written book "Creation and the Modern Christian" (Creation And The Modern Christian, Master Book Publishers, El Cajon, California, 1985) states: "If evolution were *true then the different stages of human evolution must be the best documented of all, because man is supposedly the latest evolutionary arrival, and because there are far more people researching this field than any other for fossil evidence. However, as highlighted above, the current evidence is still extremely fragmentary and highly doubtful. It is still a hotly contested matter among evolutionary anthropologists to define exactly which hominid fossils might be the ancestors of man, when, and in what order."*

H. Morris points out that the long-awaited common ancestor of man and ape, especially "Autralopithecus" including the famous "Lucy" (supposedly the oldest hominid fossil), now seems to still live in the form of a pygmy chimpanzee known as the "bonobo".

218

The "bonobo" lives in the jungles of Zaire and is almost identical to "Lucy" in body size, stature and brain size. (Science News, February 5, 1983, p. 89). The simian fable is roughly drawn in this way: First: shrews, second: common monkeys; third: Australopithecus and Ramapithecus; fourth: homo habilis; fifth: homo erectus; sixth: homo ancestor and Heidelberg man; seventh: archaic man; eighth: modern man (Homo sapiens sapiens). Let's see if this description has any solid foundation.

THE FALSE NATURE OF *Australopithecus*

The famous assumption of an "upright walk" in Australopithecus is something that has been held by paleoanthropologists such as Richard Leakey and Donald C. Johnson for decades. However, many scientists have deeply studied the structures of the Australopithecus skeletons and have proven the invalidity of that argument. Two world-renowned anatomists, Lord Solly Zuckerman of England and Professor Charles Oxnard of North America, carried out lengthy investigations on various specimens of Australopithecus and have shown that they were not bipedal and that they moved practically like modern-day monkeys. After Lord Zuckerman and his team of specialists studied the fossil bones for 15 years, with the help of the British government, they came to the conclusion that Australopithecus was only a species of common monkey, and conclusively not. he was bipedal. Note that Lord Zuckerman was an evolutionist (Solly Zuckerman, Beyond The Ivory Tower, New York: Toplinger Publications, 1970, pp. 75-94).

For his part, Charles E. Oxnard, another evolutionist known for his research on the subject, also linked the skeletal structure of Australopithecus to modern orangutans (Charles E. Oxnard, "The Place of Australopithecines in Human Evolution: Grounds for Doubt ", Nature, vol. 258, p. 389). Finally, in 1994 a team from the University of Liverpool in England undertook extensive research to come to a definite conclusion: they came to the conclusion that " *Australopithecus was quadrupedal*." (Fred Spoor, Bernard Wood, Frans Zonneveld, "Implication of Early Hominid Labryntine Morphology for Evolution of Human Bipedal Locomotion", Nature, vol 369, June 23, 1994, pp. 645-648). To put it in other words, Australopithecus have no ties to humans, they are simply a species of monkeys, just like the Ramapithecus.

The cover of National Geographic magazine with a hairy girl like a monkey, but smiling like any neighbor's daughter, went around the world when it was exposed. After this virtual reconstruction of a small Australopithecus afarensis, which lived 3.3 million years ago, there is a long story that has revolutionized world paleontology: the discovery in 2000 of the oldest fossil of a young hominid, a three-year-old girl old, in a remote region of northeastern Ethiopia called Dikika. Its skull and the bones of the upper part of its body were practically complete, which has yielded an unusual volume of information about its species. Its foot has also been found, whose analysis has certified that it was bipedal. The details of how it was found will ignite more than one vocation among those who believe that paleontology can still be a discipline tinged with adventure.

The author of the discovery, Zeresenay Alemseged (Axum, Ethiopia, 1969), has spoken about it in Barcelona invited by La Caixa Social Work (J. Á. Martos. Barcelona, May 23, 2007. ElPaís.com). In 1973, Sir Solly Zuckerman and Charles E. Oxnard presented a letter at a symposium of The Zoological Society of London. In this conclusion Zuckerman wrote: ' *Over the years I have found myself pretty much alone in challenging conventional wisdom about Australopithecus . . . but I'm afraid I've had little effect. The most authoritative voice has spoken, and its message has in due course been incorporated into textbooks throughout the world*." In this regard, he and his partner continued to defend, over and over again, the great scientific error of including Australopithecus as a hominid or "half-human" being, since it is identical to chimpanzees. On this, Oxnard added at the time: "*The conventional notion of human evolution must now be harshly modified or even rejected [...] new concepts must be explored.*" (Oxnard CE The Order of Men. New Haven, Yale University. 1984)

In 1985, Alan Walker of Johns Hopkins University discovered a fossil hominid skeleton stained by dark materials west of Lake Turkana, for which he was nicknamed the "Dark Skull." Immediately it was tried to include in the fanciful order of the hominid column that would unite the ape with the man. According to the theory that was born at that time, there would be two branches – the trunk would be the recent discovery of Australopithecus afarensis. From one of these branches would come the line called "Homo", proceeding from Homo habilis (remember that it really is a monkey), to Homo erectus (a cousin of man), to Homo sapiens. In the second branch are the australopithecines, arising from Australopithecus afarensis. According to Johanson and White, Australopithecus afarensis gave rise to Australopithecus africanus, and this in turn would have given rise to Australopithecus robustus.

Then from Autralopithecus robustus would come Australopithecus bosei. The point is that The Black Skull (KNM-WT 17000) type Australopithecus bosei was dated 2.5 million years, that is, much older than Australopithecus robustus. Johanson himself accepted that this fact complicated the relationship between all these asutralopithecines. Likewise, the general study of these fossils, including the misnamed Homo habilis, greatly complicated evolutionary theory. Pat Shipman said in 1986: "*The best answer we can give right now is that we no longer have a very clear idea of who gave rise to whom.*" (Pat Shipman. Baffling limbo in the family tree. 1986, Discover, 7 (9): 87-93)

The question of the origin of the "Homo" line remained a problem for theorists. Pat Shipman said he had seen Bill Kimbel, an associate of Donald Johanson trying to manipulate the phylogenetic implications of The Black Skull - this would not be the first time nor the only time -: «*At the end of a lecture on the evolution of Australopithecus, he erased all the order, alternative diagrams and looked at the board for a moment. Then he returned to the classroom and threw up his hands.*» Shipman wrote. After considering various phylogenetic alternatives and even finding all the evidence for them, Shipman said: " *We could say that we have no evidence even where Homo arose from and remove all members of the genus Australopithecus from the hominid family [...] I have a type of reactive visceral reaction on this idea that I suspect that I do not see myself capable of rationally evaluating. I was brought to the notion that Australopithecus was a hominid.*" Regarding this comment, archaeologist and researcher ML Cremo added: " *This is one of the most honest statements we have heard from a prestigious scientist involved in paleoanthropological research.*" (Michael Cremo, Hidden History of the Human Race, 1999. P. 265)

LUCY: A MONKEY POSING as a Human

A discovery by Donald Johanson in Hadar, Ethiopia, about an Australopithecus nearly 4 million years old has been made to sound like one of the biggest archaeological discoveries about human evolution. The biggest discoveries were made in late 1974. These remains were dubbed "Lucy" but were quickly "humanized" before the public in order to fit customarily into conventional theory. This Australopithecus afarensis tried to rebuild itself with pieces from an entire family found at the site, but to Johanson himself the skeletal reconstruction "looked *much more like a small female gorilla.*" Upon further exhaustive study it was concluded that this Australopithecus afarensis was more ape-like than human-like (Randall L., Susman, Jack T., Stern, Charles E. Oxnard).

At present, most experts say that Lucy was just an unusual chimpanzee about 3 feet tall, and not a missing link (see also: Seminar, part 7. By Dr Kent Hovind, distributed by Chick Publications).

THE HANDY MAN OR THE Handy Monkey?

His name means "skilled man" and refers to the discovery of lithic instruments probably made by him. Detailed studies of the skeletal remains of his hands have been carried out to verify whether it would really be possible for this Homo habilis to have carried them out. The scientists concluded that it was capable of performing tasks such as grasping to carry out the necessary manipulations in the manufacture of stone utensils; It was probably an opportunistic carnivore. An important increase in brain size is observed in them with respect to Australopithecus, which has been calculated between $650cm^3$ and $800cm^3$, in the crushed skull "1470", found in Koobi Fora. The remains have been found in Kenya, in the town of Koobi Fora and in Tanzania, in the well-known Olduvai Gorge. Some authors question its belonging to the "Homo" group, based on a restrictive interpretation of the diagnosis of the genus, and assign it either to Australopithecus or propose that a new genus be defined for this species in which it is also included. to Homo rudolfensis.

The story that homo habilis were transitory beings between man and ape is quite popular, and before them tools were not used among primates. The classification of "Homo Habilis" was presented in the 60's by the entire Leakey family, "fossil hunters". According to the Leakeys, this new species, which they classified as Homo habilis, had a relatively large cranial capacity, as well as the disposition to walk upright and use wooden and stone tools. Therefore, it could have been the ancestor of man. New fossils of the same species unearthed in the late 1980s changed the previous view.

Some researchers such as Bernard Wood and C. Loring Brace, who based themselves on the newly found fossils, said that Homo habilis, which means "man capable of using tools," should be classified as Australopithecus habilis, which means "South African ape capable of using tools." using tools," because Homo habilis had a lot of characteristics in common with monkeys called australopithecines. They had long arms, short legs, and a skeletal structure similar to that of australopithecines. The fingers and toes were good for climbing. The maxillary structure was very similar to that of modern monkeys.

Investigations carried out in the following years showed that Homo habilis did not have any difference with the Australopithecines. The OH62 fossil skull and skeleton found by Tim White showed that this species had a small cranial volume, long arms, and short legs that allowed it to climb trees like modern monkeys. Detailed analysis conducted by American anthropologist Holly Smith in 1994 indicated that Homo habilis was not "homo" or, in other words, not "human" but "monkey". Smith said the following about the analysis carried out on the teeth of Australopithecus, Homo habilis, Homo erectus and Homo Neanderthals: «*Circumscribed analyzes (to the teeth) of fossil specimens, exhibit patterns of their development in the graceful Australopithecus and Homo habilis that places them in the classification of African monkeys. The same analyzes on Homo erectus and Neanderthals classify them with humans.*" (Holly Smith, American Journal of Physical Anthropology, vol. 94, 1994, pp. 307-325).

That same year, Fred Spoor, Bernard Wood and Frans Zooneveld, all specialists in anatomy, reached the same conclusions through a totally different method that was based on the comparative analysis of the semicircular canals of the inner ear of humans and of the monkeys. These channels are related to balance. The channels of humans, who walk upright, differed considerably from those of monkeys, which walked bent forward. The inner ear canals of all Australopithecus and Homo habilis specimens analyzed by Spoon, Wood, and Zooneveld were the same as those of modern monkeys. The inner ear canals of Homo erectus were the same as those of modern man (Fred Spoor, Bernard Wood, Frans Zonneveld, "Implication of Early Hominid Labryntine Morphology for Evolution of Human Bipedal Locomotion", Nature, vol 369, June 23, 1994. Pages 645-648).

This discovery leads to the important conclusions that the fossils of Homo habilis did not really belong to the class "homo", that is, to the human class, but to the Australopithecine class, that is, to that of monkeys. In addition, Homo habilis, like Australopithecines, were living beings that walked bent over and striding, so their skeletons were monkeys. They had no relationship of any kind with humans. Based on the skeletons found, it is believed that this type of chimpanzee lived approx. from 2.5 to about 1.44 million years before present (at the beginning of the Pleistocene). And it is "funny", because there is no other way to say it, but the classification of "intelligent beings capable of using tools" should go back to 2.8 billion years, that is, when the development of the first forms of life began to be seen complex on our planet.

Interestingly 5 years after the discovery of Mono-habilis, Bryan Patterson and WW Howells found modern hominin humeri in Kanapoi, Kenya. In 1977, French workers found similar humeri in Gombore, Ethiopia. The age of these is around 4.5 million years. This humerus found by Patterson and Howells was very different from that of gorillas, chimpanzees and australopithecines, and rather similar to humans.

So that the many current discoveries leave these inconsistent postulates very inconsistent. We have been wrongly taught that there has been a human evolutionary order that came from Australopithecus (Australopithecus afarensis, africanus and robustus. 3'000,000-500,000 years) in Africa, then came Homo Habilis discovered in Olduvai Gorge in Tanzania. But the tools of this "skillful" ape were dated to 1,750,000 years ago, so it would be interesting to see how both species met halfway through their evolution. This is so because there is evidence from the 80's where Australopithecus are seen with an unknown type of "man" ("The Prehistoric World" by MacDonald Educational, Ediciones Vidorama, Barcelona - Spain. 1987), so that these species came to live in East Africa 2.5-3 million years ago.

HOMO RUDOLFENSIS: A Badly Assembled Face

This species was proposed by Valerii P. Alexeev in 1986, whose type specimen is KNM-ER 1470, found at Koobi Fora (eastern shore of Lake Turkana, formerly Lake Rodolfo), by Bernard Ngeneo, a member of Richard Leakey's team. , in 1972. Alexeev designated it in 1986 as Pithecanthropus rudolfensis, although it has subsequently been assigned to both the Homo and Australopithecus genera. Some authors propose that a new genus be defined for this species in which Homo habilis is also included. Some paleoanthropologists doubt that it is a different species from Homo habilis, but this is the prevailing opinion at present, due to marked morphological differences, among which the following must be distinguished: shape of the face (mainly in the supraorbital region and malar, which is very long, deep and inclined forwards); cranial measurements as a whole (45% of the measurements that were compared between the two species exceed the sexual dimorphism of gorillas) and cranial volume (around 750cc, compared to 500cc for Homo habilis), although in 2007 the The cranial capacity of Homo rudolfensis has been estimated by Timothy Bromage, an anthropologist at New York University at 526cc.

Likewise, anatomically, Homo rudolfensis has, compared to Homo habilis, a flatter face, broader post-canine teeth with more complex roots and crowns, and thicker enamel. Richard Leakey, who unearthed the fragments, presented the skull - dubbed "KNM-ER 1470" and considered to be 2.8 million years old - and said it was the greatest discovery in the history of anthropology. with sweeping effect. According to Leakey, this being, which had a small cranial volume like the Australopithecus, and yet a human face, was the missing link between the Australopithecus and the human being. Even so, a short time later it was understood that the "human type" face of the KNM-ER 1470 skull, which frequently appeared on the cover of scientific journals, was the result of an abnormal assembly of the found fragments, something that could be deliberate.

Professor Tim Bromage, who did studies on the anatomy of the face, underlined this fact that he discovered with the help of computer simulation in 1992: «*When (the KNM-ER 1470) was first rebuilt, the forehead was fitted to the skull. in a nearly upright position, much like that exhibited by flat modern human faces. But recent studies of the anatomical relationships show that in life the face must have protruded considerably, giving it a monkey-like appearance, like the faces of Australopithecus.*" (Tim Bromage, New Scientist, vol. 133, 1992, pp. 38-41). The evolutionary paleoanthropologist JE Cronin has the following to say about it: «... *its vigorous rostrum, the flattened naso-alveolar clivus (reminiscent of the concave rostrums of australopithecines), the reduced maximum cranial breadth (in the temporal bones), the pronounced canine and the large molars (as indicated by the remains of the roots), are all relatively primitive features that relate the specimen to members of the taxon (genus) Australopithecus africanus .*" (JE Cronin, NT Boaz, CB Stringer, Y.

Rak, "Tempo and Mode in Hominid Evolution", Nature, vol. 292, 1981, pp. 113-122).

C. Lorng Brace of the University of Michigan came to the same conclusion as a result of his analyzes of the jaw and tooth structure of the 1470 skull and said that *"the size of the jaw and the part containing the molars showed that ER 1470 had exactly the face and teeth of the Australopithecus."* (CL Brace, H. Nelson, N. Korn, ML Brace, Atlas of Human Evolution, 2.b. New York: Rinehart and Wilson, 1979). Professor Alan Walker, a paleoanthropologist at John Hopkins University, who had done as much research as Leakey, maintains that this living being should not be classified as "homo" - referring to Homo habilis or Homo rudolfensis - but, on the contrary, should be classified as "homo". must be included among the members of the Australopithecus species. (Alan Walker, Scientific American, vol. 239 (2), 1978, p. 54). Ultimately, the qualifications as Homo habilis or Homo rudolfensis, presented as transitory links between Australopithecines and Homo erectus, are totally imaginary. As has been confirmed by many researchers today, these living things are members of the Australopithecus series. All their anatomical features reveal that both are species of monkeys.

JAVA MAN (PITHECANTROPUS erectus)

The first specimens of Homo erectus were found on the island of Java in 1891 by Eugène Dubois. Java man was initially named Pithecanthropus erectus, but was later transferred to the genus Homo. This man from Java (Homo erectus erectus) was the first representative of Homo erectus to be discovered. The word "pithecantropus" derives from Greek roots and means "ape man". Dubois found the remains at the Trinil site (Java Island) in 1891. What makes the discovery unreliable is that this finding only consisted of the cover of a single skull. A year later a femur and two molars were discovered 16m from where the skull cap was found, which implies that it is highly unlikely that they belonged to the same individual. However, Dubois tendentiously considered that all the pieces came from the same being and dated them as half a million years old.

Ergo, he did not reveal, until thirty-one years later, that he had also found two obviously human skulls at the same date and place. Most of the evolutionists of those days were convinced of the validity of this half-million-year-old creature, and misinformation about it was growing. In 1940 Weidenreich reinterpreted the remains as Homo erectus iavanensis, but it was definitively renamed by Dobzhansky (1944) as Homo erectus erectus. It had a cranial capacity of about 940ml, intermediate between the 1,200 to 1,500 of modern man and the 600ml of the gorilla. Java Man possessed the portion of the brain that controls language, although whether he actually spoke is unknown. The Java man brain was much larger and more convoluted than that of any primitive or living ape, and had more human than apelike features.

The adult was about 1.70 m tall and weighed about 70 kg, in addition to walking upright. It is assumed that the men of Java possibly moved in small family groups, lived in caves and hunted in the forests. According to evolutionary theory, Asian Homo erectus would have come from African Homo erectus since the accepted hypothesis says that our ancestors arose from there, and from there they migrated to other locations. However, on the Kabu formation in Trini, where Dubua found its original Java man, the soil has been studied with potassium-argon dating and has given 800,000 years. Other discoveries in Java come from the Djetis layers in the Putjangan formation.

According to T. Jacob, these layers near Modjokerto at the convergence of the Lower Pleistocene dated with potassium-argon at 1.9 million years. The important thing about this is that African

Homo erectus would not have migrated out of Africa until a million years ago. In fact, African Homo erectus were dated to 1.6 million years, which would place the "supposed" ancestor of man from Asia and not from Africa. Another point that overturns the accepted hypothesis. Other studies carried out by MH Day and TI Molleson in 1973 (Symposium for the Study of Human Biology, 2: 127-154) would determine, thanks to the "fluorine to phosphate" theses -originated by KP Oakley-, that the femur of Java (Pitecantropus erectus) would be about 800,000 years old. So the 1.9-million-year-old cap of the skull would possibly be of a monkey type, while the femur would be of an 800,000-year-old human, again ruling out this most important specimen for evolutionary theory as one of the earliest. and more significant human transition beings.

PEKING MAN (SINANTHROPUS)

Also called "Homo erectus pekinensis" or "sinanthropus pekinensis", this primate is considered a subspecies of Homo erectus typical of China. Its name alludes to the fact that its fossil remains were discovered southwest of Beijing, in a cave in the town of Zhoukoudian between 1921 and 1937, and date from between 250,000 and 500,000 years ago. It is especially popular because at the time of its discovery it was considered the first "missing link" justifying the theory of evolution. According to Wikipedia, for years the inhabitants of the area sold all sorts of strange or old-looking teeth to foreigners, pretending they were dragon teeth, and chance arose when one of these teeth ended up in the hands of a Swedish scientist, who, upon studying it, recognized it as belonging to an extinct mammal.

The origin of that tooth was investigated and it was established that it came from a cave in Beijing. Investigations began in 1921. According to the later account of Otto Zdansky, who worked for geologist Gohan Anderson, a local resident led archaeologists to what is now known as Dragon Bone Hill, a place full of fossilized bones. Zdansky began his own excavation and eventually found bones that looked like human molars. In 1926 he took them to the Peking Medical College, where the anatomist Davidson Black analyzed them. Later, he would publish his discovery in the journal Nature. Of course, the Rockefeller Foundation agreed to sponsor the work at Zhoukodian, clearly being the Rockefeller billionaires the largest Masonic, atheist and anti-religious organization on our planet.

Around 1929, the Chinese archaeologists Yang Zhongjian and Pei Wenzhong, and later Jia Lanpo, took charge of the excavation. Over the next seven years they unearthed fossils of more than forty specimens of adults, juveniles, and children, including six nearly complete cranial vaults. It is believed that the place was a burial site. The paleontologist Pierre Teilhard de Chardin and the anthropologist Franz Weidenreich also participated in the discoveries. The excavations ended in July 1937, when the Japanese occupied Beijing during the Second Sino-Japanese War. The fossils were kept safe in the Cenozoic Laboratory of the Faculty of Medicine. In November 1941, Secretary Hu Chengzi sent them to the United States to protect them from the impending Japanese invasion. However, on the way to the port city of Qinghuangdao, they disappeared, allegedly at the hands of a group of marines that the Japanese had captured at the start of the war with the US.

WHAT WAS PEKING MAN like?

Due to the disappearance of the fossil remains, subsequent researchers have only been able to count on the molds and writings made by the discoverers. Thus, it is known that its cranial capacity reached 1075cc, 80% compared to that of Homo sapiens, and that it was a hunter-gatherer. The discovery of animal remains along with the bones and the evidence of the use of fire, to fight the cold and to cook food, and of bone and wood tools, made with others of stone, served to support the theory that the Herectus was the first faber species. The analyzes led to the conclusion that the Zhoukoudian and Javan fossils belong to the same stage of human evolution. This is also the official view of the Communist Party of China. However, this interpretation changed in 1985 when Lewis Binford stated that the Peking man (Sinanthropus erectus) was not a hunter, but a scavenger. In 1998, Steve Weirner's team at the Weizmann Institute of Science concluded that there is no evidence that Peking Man used fire.

HOMO ERECTUS, THE HUMAN who has been made to pass as a Monkey

Another of the supposed links in the evolutionary chain, and the most important is Homo Erectus. The fossils that have made it known in the world are those of Peking Man and Java Man found in Asia. In any case, it was realized over time that both fossils were not trustworthy. The Peking Man consisted of some elements made of plaster, the originals of which have been lost. Java Man was "composed" of a skull fragment and a pelvic bone found a few meters away from the first, without any indication that they belonged to the same living being. That's because Homo erectus fossils found in Africa gained increasing importance. (It should be noted that some of the fossils said to be Homo erectus were listed by some evolutionists under a second class called "Homo ergaster". There is disagreement among them on this. To make things easier, better call all these fossils under the classification of Homo erectus).

The most famous human "link" in the evolutionary legend is Homo erectus, of which there is little reliability when it comes to relating it to Homo habilis and although it also lived in Africa, it has been discovered in what was then Eurasia. However, there was a great space of more than 250,000 years until it came to appear (1'800,000 - 500,000 years approx.) if it is associated with the evolutionary chain of man. It is believed that a type of Homo erectus whose evidence dated from 500,000 to 250,000, could be a human link, but it was some kind of creature that resembled a Neanderthal in the shape of the skull, which has made it the most credible for the next "transitory being". Then, in the last European inter-glaciation (200,000 - 100,000 years) other evidence seems to have appeared, because the Neanderthal species suddenly appears and in the same way is "eradicated" overnight to make way for Homo sapiens sapiens.

Some believe that the first hominids to come out of Africa lived in Georgia 1.8 million years ago, and were like "Mzia" discovered by David Lordkipanidze, Rolex Prize (Very Interesting - no. 283 Dec. 2004), but about 2 million years ago. years there was a human being equal to the current ones who left a skeleton in Castenodolo in Italy -later we will emphasize this topic-, which proves the truth of the Nazi Hoerbiger that there were men in the Tertiary (65 million years ago), with the dinosaurs, when official science tells us that we started in the Quaternary, that is, about 1.6 million years ago. Although other evidence found in Egypt with 3 million years put into question once again the evolutionary theories. Professor William Laughlin of the University of Connecticut conducted extensive anatomical examinations of the Eskimos and people living on the Aleutian Islands and noted that they were remarkably similar to Homo erectus. Laughlin concluded that all these races were actually different varieties of Homo Sapiens (modern man).

When we consider the vast differences that exist between widely separated groups such as the Eskimos and the Bushmen who belong to the same species of Homo sapiens, it seems justifiable to conclude that the synanthrope (a class of erectus) is included in that same variety. (Marvin Lubenow, Bones of Contention, Grand Rapids, Baker, 1992. p. 136). « *There is a great gap between Homo erectus, a human race, and the monkeys that preceded it in the scenario of the theoretical "human evolution" (Australopithecus, Homo habilis, Homo rudolfensis). This means that the first men appeared in the fossil record suddenly and immediately without any evolutionary history. There can be no clearer indication that they were created.* » (Harun Yahya in "The Deception of Evolutionism"). However, admitting this fact goes totally against the dogmatic philosophy and ideology of evolutionists. Consequently, they try to portray Homo Erectus, truly a human race, as a half-simian race, as well as the rest of the skeletons.

In their reconstructions of Homo erectus, they stubbornly drew it with simian features, creating a fabled vision in society. On the other hand, with similar drawing methods, they humanized monkeys such as Australopithecus or Homo Habilis, since they are only types of chimpanzees. With this procedure they seek to "approximate" monkeys and human beings and close the gap between these two different classes of living beings within an imaginary world of evolutionary transitions that never existed. So the famous fossil of the Turkana Child, which belonged to the Homo erectus race, hardly differed from us. All "homo" classifications or types actually include original human races (not distinct species). "*The difference between them is no greater than the difference between an Eskimo and a Negro or a Pygmy and a European.*" (Harun Yahya, "The Deception of Evolutionism")

Homo erectus, which is referred to as the most primitive human species, means: "man who walks upright." Evolutionists have had to separate these men from earlier ones by adding the quality "upright," because all available Homo erectus fossils are upright to a degree not observed in any of the Australopithecine or Homo Habilis species. Let's keep in mind that there is no difference between the skeleton of the modern human being and Homo Erectus. The primary reason for evolutionists to define Homo Erectus as "primitive" is the volume of the skull (900-1000cc) - smaller than the average for modern humans - and the protrusion of the brow ridge. However, many people living today have the same cranial volume as Homo erectus (eg, pygmies), and there are also races with protruding browbones (eg, Australian aborigines).

A problem in these evolutionary speculations appears when they find earlier human evidence, such as the case of the entirely human skeleton found in Olduvai Gorge, Tanzania. These bones found in 1913 by H. Reck are 1.15 million years old, an impossible estimate for a modern man, and even less so in this region where Homo erectus are supposed to have lived. Recent discoveries by paleoanthropologists have revealed that certain segments of humans classified as Homo erectus have lived until very recently. Neanderthalian Homo sapiens and Homo sapiens sapiens (modern human) coexisted in the same region. This situation apparently indicates the invalidity of the assumption that one is the ancestor of the other. It is more difficult to justify how the Neanderthal man disappeared overnight and Homo sapiens appeared, too, overnight.

HOMO ERGASTER, ANOTHER "Erectus"

The best known class of Homo erectus was found in Africa and corresponds to the "Narikotome homo erectus" or "Turkana Boy" fossil, which was found near Lake Turkama in Kenya. It was confirmed that this fossil was of a 12-year-old boy who would have been 1.83m tall in adolescence, quite enough to be an ancestor of man. The vertical structure of the fossil skeleton does not differ in any way from that of modern man. The American paleontologist Alan Walker said regarding the same that he doubted that *"the average term of pathologists could say what were the differences between that fossil skeleton and the skeleton of the modern human."* (Boyce

Rensberger, The Washington Post, November 19, 1984). Regarding the skull, Walker said that *it "looked just like a Neanderthal."* Neanderthals are a modern human race.

Therefore, Homo erectus is also a modern human race, and in that same order, then also Java and Peking man. Even the evolutionist Richard Leakey says that the differences between Homo erectus and modern man are nothing more than racial variations: *"One should also look at the differences in the shapes of the skull, in the degree of protrusion of the face, in the vigor of the eyebrows etc These differences are probably no more pronounced than those we see today between geographically distant human races. Such biological variations arise when populations are geographically separated for a significant amount of time.»* (Richard Leakey, The Making of Mankind, London: Sphere Books, 1981, p. 62).

Homo ergaster is a type of Homo erectus native to Africa. It is estimated that it lived between 1.75 and a million years ago, in the Calabriense (middle Pleistocene). Its first remains were found in 1975 in Koobi Fora (Kenya); These are at least two skulls (KNM-ER 3733, perhaps female, and KNM-ER 3883) from 1.75 million years ago whose brain had an estimated size of about 850cm^3. Then, in 1984, the complete skeleton of a boy of about 11-12 years of age, 1.60m tall and a brain of 880cm^3, with an age of 1.6 million years, was discovered in Nariokotome, near Lake Turkana (Kenya). years. This is what is known as the child of Nariokotome, and it does not differ practically nothing from a current child. Some specialists consider that they may have been a single species, due to their great anatomical resemblance, in which case their name as Homo erectus would have priority, but the acceptance of two different species seems to be established in some cases.

This statement is incorrect since they are simply humans from the same family but with natural differences that beings of the same race have. But this is not always noticed or taken into account by evolutionists, who have made some of the fossils that were said to be Homo Erectus, part of a second class called "Homo Ergaster". On this there are many notable disagreements. In other words, all these fossils should be called under the classification of Homo Erectus, just as today all of us are called human beings, even though we differ in skin color, eye color, hair type, and eye shape, height or bone complexion.

On one occasion something shocking was discovered. They are naval engineers who are 700,000 years old, that is, ancient mariners: " *The first humans were much more ingenious than we suspected ...*" News published in New Scientist on March 14, 1998 tells us that humans called Homo erectus by evolutionists were professional sailors 700,000 years ago. Those humans, who had enough knowledge and technology to build a ship and who had a civilization that used shipping, can hardly be called "primitive." In short, in his book Hidden History of the Human Race, its writer, Michael Cremo, highlights the most important points that are overlooked in the evolutionary history from ape to man:

1. There is a significant body of evidence from Africa suggesting that humans anatomically identical to us were present in the Early Pleistocene and Pliocene.
2. The controversial image of Australopithecus as a true terrestrial semi-human biped proves to be false.
3. The status of Australopithecus and Homo erectus as human ancestors is questionable.
4. The status of Homo habilis as a distinct species is questionable.
5. Even confining ourselves to accepted conventional evidence, the multiplicity of proposed evolutionary connections to hominids in Africa paints a very confusing picture. Combining these findings with those of the

previous chapters, we conclude that all the evidence, including fossilized bones and artifacts, is most consistent with the view that anatomically modern humans coexisted with other primates for tens of millions of years.

Homo Antecessor

Homo antecessor is the name given to some Homo erectus bones over a million years old (Lower Pleistocene) found in Spain. The definition of this species is the result of the more than eighty remains found since 1994 at level TD6 of the Gran Dolina site in the Sierra de Atapuerca. A very well preserved mandible of a female Homo antecessor, between 15 and 16 years old, recovered from the La Gran Dolina site has very clear similarities with those of Peking Man (Homo erectus), suggesting an Asian origin of Homo antecessor. However, the pattern of tooth development and eruption is virtually identical to that of modern populations.

At present, the validity of this denomination as a different species is defended by its discoverers and other experts, who consider that Homo antecessor precedes Homo heidelbergensis and therefore is also an ancestor of Homo neanderthalensis; However, part of the scientific community considers it a simple, non-specific name to refer to remains found in Atapuerca, which they assign to the species Homo heidelbergensis or, they consider it a variety of Homo erectus / Homo ergaster. In March 2008, new remains of Homo antecessor were revealed, specifically part of the jaw of an individual about 20 years old and 32 Olduvayense-type flint tools, dated to 1.2 million years old, which makes considerably regress the presence of hominids in Europe.

The remains were found in 2007 in 'La Sima del Elefante', a site located about 200 m from 'La Gran Dolina'. In this order of so many mysteries, how did Homo Antecessor himself get to Atapuerca? Nobody knows. They assume - it is "one more hypothesis" - that it arrived by land from the East, not by the Strait. The fossils of Atapuerca (Spain) are 800,000 years old, and there are even some up to 1.5 million years old. Ergo, paleontologists attribute that what they have discovered is not a primitive Homo Antecessor, but a Homo Ergaster or a Georgic Homo. Although Juan Luis Arsuaga himself assumes: «*human presence there has been a million years or almost*». What are we left with then? Neither the scientists themselves have the answers, or they have them and "they don't want to see them!"

THE CHILD OF LA GRAN *Dolina*

One of the human fossils that has aroused the most interest was found in Spain in 1995. The fossil in question was found by three Spanish paleontologists from the University of Madrid. It was discovered in a cave called Gran Dolina in the Atapuerca region. The fossil revealed the face of an 11-year-old boy identical to a modern man, however the boy had died 800,000 years ago. This fossil made even Juan Luis Arzuaga Ferreras, who directed the excavation of Gran Dolina, doubt. Here's what Ferreras said: «*What we found was a completely modern face, for me, this is the most spectacular, it's the kind of thing that throws you off, finding something as unexpected as that. But the most spectacular thing is to find in the past something that you think belongs to the present. It would be like finding a tape recorder in Gran Dolina. This would be very surprising, we do not expect to find cassettes and recorders in the lower Pleistocene. Finding a modern face from 800,000 years ago is the same thing.*"

THE MAN FROM HEIDELBERG

This other supposed human link is a representation based on a jaw that many experts consider to be simply a human jaw, and not something related to a human ancestor. This would be a direct ancestor of the Neanderthal Man in Europe, under the evolutionary prism, given his resemblance to the Neanderthal Homo sapiens; even when it is very similar to archaic Homo sapiens, found in Africa, such as Homo rhodesiensis and Homo sapiens idaltu, according to some defenders; it is known today that Homo heidelbergensis was not a direct ancestor of modern humans. Between Homo antecessor, whose fossils have been found in the hills of Atapuerca (Spain), and Homo neanderthalensis, there was this species (Homo heidelbergensis). It generally presents characters intermediate between Homo erectus/Ergaster and Homo sapiens, including a cleft occipital torus (in Homo erectus such a torus, or crest, is continuous) and a large neurocranial capacity.

The oldest fossil of the species is a lower jaw found by a mine worker at Mauer, near Heidelberg. Later, in a cave called Caune de l'Arago, in France, the fragmentary remains of a dozen individuals were found. The most complete is the face and part of the braincase of an individual known as Tautavel Man, which bears a strong resemblance to the Petralona Man skull found in a cave in Greece. Other sites where fossils of this species have been found are Steiheim (Germany), Swascombe (England) and the Sima de los Huesos in 'La Sierra de Atapuerca' (Spain), where 5,000 fossils belonging to about 30 individuals were found, dating back to from 400,000 years ago and more, considered ancestors of the Neanderthals, remains which are very well preserved.

Among them, the skull number 5 (popularly called "Miguelón") stands out, which is complete, and of which studies have recently been carried out that show a laterality in the brain (he was right-handed), and a very well preserved pelvis of a popularly known individual. like "Elvis". Fossils consistent with this group have been found in China at the Dali site; a skull from 280,000 years ago, and a skeleton at Jinniushan.

In 1972, David Pilbeam said of the Heilderberg mandible that it " *seems to date from the Mindel glaciation, and its age is sometime between 250,000 and 450,000 years.*" The German anthropologist Johannes Ranke wrote in the 1920s that the Heidelberg mandible belonged to a representative Homo sapiens rather than a

242

monkey-like half-ape ancestor. According to Frank E. Poirier (1977), the teeth of the Heidelberg mandible are more similar in size to those of modern Homo sapiens than to Asian Homo erectus (Java and Peking men) this was also written about by TW Phenice of Michigan State University in 1972: " *The teeth are remarkably equal to those of modern man in almost every respect, including size and relief pattern.*" (Michael Cremo, The Hidden History of the Human Race, 1999, p. 163)

HOMO RHODESIENSIS

Homo rhodesiensis was found for the first time in 1921 in the locality called Broken Hill by the English, currently Kabwe, in Zambia (formerly "Northern Rhodesia" for which reason it was called Rhodesian Man). It is considered that it lived only in Africa, from 600,000 to 160,000 years before present, during the Ionian (middle Pleistocene). In this same area of Broken Hill that year, in those excavations, a Neanderthal man was discovered with a hole in his skull, the product of a bullet shot. What was impressive was that it was found at a depth of 18 m, which estimates its age at 40,000 years. Forgetting for a moment the fact that 40,000 years ago no one is known to have carried weapons, the fact that a Neanderthal was found in the same area and with the same age implies that they were from the same family.

Phillip Rightmire (1998) considers that African fossils from the Middle Pleistocene should be included within the species Homo heidelbergensis, from which both Neanderthals and Homo sapiens would therefore descend – what we now understand -: they were relatives, cousins. For his part, the French paleoanthropologist Jean-Jacques Hublin (2001) assumed that Homo rhodesiensis is a precursor species to Neanderthal Man, which had no role in the phylogenesis of Homo sapiens. But, according to Tim White (2003), it is very likely that Homo rhodesiensis is an ancestor of Homo sapiens idaltu. If the evolutionists themselves don't know where to put the human skeletons, why do they teach us unsupported concepts? All they try to do is forcefully search for transitional beings, even though they don't exist. *"We must turn to the fossil record to find an answer to the question of when man appeared on Earth."*

This record shows that man appeared millions of years ago, and not thousands ago, as the popular hypothesis defends. These discoveries consist of skeletons and skulls, and the remains of people who lived at different times. One of the oldest traces of man are the prehistoric "footprints" found by the famous paleontologist Mary Leakey in 1977 in the Laetoli region of Tanzania. Research indicates that these footprints were 3.5 million years old. Much older than any kind of hominid. Proof of an evolutionary sequence requires at least one of two types of evidence: either an unbroken chain of surviving transitional or intermediate fossils, or plausible reconstructions of such series together with their respective ecological niches.

The difficulty is to show how each link in the chain could be viable long enough for the next to establish itself. Only by establishing complete transition series can the hypothetical continuity of the hierarchy be made plausible—of course, empirical proof is a much more difficult requirement to meet. What is at issue here is mere plausibility. If such transitions never took place, intermediate forms should be found in fossils and in living organisms. Existing classes should overlap. Clearly marked boundaries should be the exception rather than the rule.

The Neanderthal: Cousin of Homo Sapiens

Modern science does not have an explanation for this fact: it is not known how one day the Neanderthals disappeared and Homo Sapiens sapiens appeared. At first it was believed that the current man appeared 40,000 years ago, however, first the archaeologist Jean Steen-Mackintyre discovered evidence in Mexico from 250,000 years ago, and later, other researchers found other evidence from 300,000 years ago in Siberia. Despite the fact that by then the archaeologist Jean Steen-Mackintyre had already been ridiculed, discriminated against and all doors had been closed to her. This evidence showed that the Cro-Magnon (first group of European men) was later than Homo erectus himself discovered in Europe.

Let's always start from the fact that it was believed that Homo sapiens had emerged 40,000 years ago, after it was 250,000 years ago, evolving from Neanderthals, but then there is evidence of men from 1.8 million years ago, a time when Homo erectus also appeared. So, Homo sapiens are not predecessors of Neanderthals or Homo erectus. Who, then, is their predecessor? For many years the Neanderthal man was considered a missing link. He was represented as a hairy, semi-erect creature with a circular chest, and most of the time with a club in his hand. Other Neanderthal skeletons revealed that Neanderthal man was fully erect, fully human, and with a brain capacity that exceeds that of modern man by 16%. It was concluded that the initial specimen was crippled by bone arthritis and rickets. Today Neanderthal Man is considered as Homo sapiens.

At the International Congress of Zoology (1958) Dr. AJE Cave said that his examination of this famous skeleton, found in France more than 50 years ago, showed it to be that of "an old *man suffering from arthritis*." Homo Neanderthals are said to have evolved in Europe from populations of Homo heidelbergensis which already displayed robust bones and deep attachment points for muscles. However, Neanderthals are not currently considered the ancestors of

Homo sapiens, but rather a species that evolved separately in Europe during the Pleistocene.

Thanks to the discovery of a Neanderthal infant from 29,000 years ago (an age close to the end of this hominid species that occurred about 30,000 years ago), researchers from the Center Human Identification of the University of Glasgow analyzed the DNA of the bones extracted from one of the the knees of this skeleton and concluded that Neanderthals contributed little or nothing at the genetic level to the evolution of Homo sapiens, which implies that they are not ancestors of modern humans.

NEANDERTHAL DNA (NEANDERTHAL Homo sapiens)

On the other hand, in 1977, the geneticist Suante Päävo, from the University of Munich, took an arm fragment from the fossilized bones that are kept as a state secret, and studied the DNA of Neanderthal man for the first time. The comparison of the genetic heritage revealed clear differences between primitive man and modern man (Homo sapiens sapiens). To get to the point, 27 differences were located in the mitochondria, while in all the breeds that currently exist, a maximum of 8 differences are observed. The explanation for that "missing link" can only be Extraterrestrial or Divine intervention. The more we learn about the Neanderthal, the less primitive it becomes. Recent scientific articles have admitted that modern man and Neanderthal met, interacted, and even interbred.

Another recent article suggests that Europeans could be 5% Neanderthals. On the other hand, the dominant paradigm of many scientists continues to insist that Neanderthals could not speak and that there was no question of interbreeding. So what would happen when scientists were able to isolate Neanderthal DNA, as recently found in material found in a Croatian cave? Neanderthal DNA is

99.9% identical to "human" DNA. A 99.9% is certainly a given that is amazingly identical, even if it is not 100% identical. What the scientists don't say is the following quote from a lecture by Eric Lanser, PhD: "*The two human beings on Earth are 99.9% identical in their DNA sequences.*" This kind of data puts Neanderthal DNA in a new light. Their DNA differs from ours, just as our DNA differs from one person to another.

That is, the Neanderthal is exactly as different from you as you are from your neighbor, only he is usually depicted hunched over, wearing animal skins, and holding a spear. (This is also brought up in an article by Fiona MacRae on November 15, 2006) Both modern men and Neanderthals coexisted for many thousands of years, before the latter died out around 30,000 years ago, perhaps beaten by their more innovative cousins. in the race for food, clothing and housing. Dr Svante Pääbo (CORR), from the Max Planck Institute in Leipzig, Germany, said: "*While it cannot be definitively concluded that interbreeding between the two species of humans did not occur, analysis of Neanderthal nuclear DNA suggests the probability of a casualty occurring at any level.*"

— ◦◦◦ —

OTHER SIMILARITIES and Differences

The international team of scientists that has prepared the first draft of the Neanderthal genome has discovered that all non-African humans only share between 1% and 4% of our DNA with that extinct man. According to the authors of the study, which is published in the journal Science, that portion of the genome is proof of hybridization that occurred shortly after the first Homo sapiens left Africa, if we accept popular legend. According to the investigations, the theory that is ventilated is that between 50,000 and 80,000 years ago, a group of humans found themselves in the Near East or the Middle East with Neanderthal populations, and both humanities mixed. When our ancestors later multiplied, divided and spread across Eurasia, they already carried Neanderthal material in their genome. Neanderthals were shorter and stockier than we are, and skilled tool makers as well.

According to some, they disappeared about 27,000 years ago, after the arrival on the continent of Homo sapiens, according to theorists, from Africa. The first Neanderthal remains were found in the Belgian cave of Engis in 1829; but the species was not named until 1857, after the discovery of part of a skull and other pieces in the Neander valley (Germany). In "Uriel's Machine," authors Christopher Knight and Robert Lomas write: "... *Neanderthals did not contribute mitochondrial DNA to modern humans; Neanderthals are not our ancestors ...*» To obtain the necessary DNA for sequencing the Neanderthal genome, an international team led by Svante Pääbo, from the Max Planck Institute for Evolutionary Anthropology, has used remains of this hominid from the Vindija sites (Croatia), Mezmaiskaya (Russia), Feldhofer (Germany) and the cave of El Sidrón (Asturias, Spain).

The preliminary analysis of the sequence and its comparison with five current human genomes – a South African San, a Yoruba, a Han Chinese, a Frenchman and a native of Papua New Guinea – have allowed the identification of 83 different genes between Neanderthals and us, and discover that there was hybridization between both species, a phenomenon that did not affect the Homo sapiens that remained in Africa. "*Non-African humans carry Neanderthal DNA in at least 10 of the 23 chromosomes,*" says Carles Lalueza-Fox, paleogeneticist at Pompeu Fabra University, co-director of the El Sidrón project and one of the co-authors of the study. «*The presence of material from this hominid in similar proportions in present-day European, Asian and Oceanic populations, and its absence in African ones, suggests that the sexual episode had to take place shortly after the ancestors of all non-Africans left the continent and before that population experienced a population explosion and spread across Eurasia."*

«It was a genetic exchange not very intense; but, as the group of Homo sapiens involved was very small and expanding, it had an impact on the entire population," explains another of the authors of the research, paleontologist Antonio Rosas, from the National Museum of Natural Sciences and also co-director of the excavations of El Sidrón. And, therefore, as Pääbo indicates, that relative that we thought was dead is not completely dead. *"In a sense, Neanderthals are not extinct. They live in some of us"*, recalls the Swedish paleogeneticist. The differences between modern humans and Neanderthals are located in 83 genes related to cognitive functions; physiology and anatomy of the skin; and skeletal development, especially of the skull. (Source: publico.es, May 6, 2010)

Pääbo and his team verified in 2010 that Neanderthals share with man modifications in the FOXP2 gene, which is related to the ability to speak: *«There are no reasons to think that they could not articulate in the way that we do»*, pointed out the scientist as he concluded somewhat skeptically: *«But that does not mean that it had a language as we know it today, since speech is the result of an infinite number of factors that do not depend on a single gene.»* In short, we can dare to say that Neanderthals and Homo sapiens were human cousins who lived together, mixed, interbred, developed the same languages and the same technology. That means that they are neither archaic nor are they evolutionary transitions.

———— ⌇ ————

THE CRO-MAGNON, AN ET among Us?

One of the oldest and best established fossils (Cro-Magnon) possesses at least the physique and brain similar to modern man. So what is the difference? Cro-Magnon Man is the name by which the human type corresponding to certain fossils of Homo sapiens (that is, the current human species) is usually designated, especially those associated with the caves in Europe where paintings were found. cave. It is usually Spanish and abbreviated as "Cromañón", especially for its use in the plural (cromañones). Cro-Magnon is the local denomination of a French cave in which the fossils from which the group was typified were found.

Its dating (40,000 and 10,000 years old) is taken as the milestone that begins the Upper Paleolithic from the anthropological point of view, while the modern limit does not mark it, the appearance of any physical modification, but environmental and cultural: the end of the last ice age and the beginning of the current interglacial period (Holocene geological period), with the cultural periods called Mesolithic and Neolithic. Until recently, the first modern European men were grouped into two varieties: the more robust Cro-Magnon race, and the more graceful Combe Capel, Brno or Predmost variety. In reality, this dichotomy was intended to justify the Aurignacian-Perigordian cultural binomial and today it has been abandoned, the use of the term Cro-Magnons being only generalized for "modern Paleolithic men". Later varieties (Grimaldi or Chancelade man) do not seem to have somatic differences that justify a complete population differentiation of racial type.

However, for a long time the erroneous identification of these three human types with the three racial divisions or human races of classical anthropology was popularized: Cro-Magnon with the white or Caucasian race, Grimaldi with the black or Negroid race and Chancelade with the Eskimos or Yellow or Mongoloid race. Geologist Louis Lartet discovered the first five skeletons in March 1868 in the Cro-Magnon cave (near Les Eyzies de Tayac-Sireuil, Dordogne, France), from which the Cro-Magnon and chancelade get their names. Cro-Magnon Man's brain appeared with certain mysteriously enhanced skeletal features and with a cranial capacity that is 100cc larger than that of modern man. In this skull we see a great "brain expansion" which has not occurred in any other species on Earth, in all ages of the past. This man was super gifted. Was he human?

From the study by Broca, Quatrefages, Hamy and Lartet of the remains of the Cro-Magnon cave (three adult males, a female and a fetus) a description was derived that included as prominent features a high height -one of the males measured 1 80 m-, prominent chin and large cranial capacity (1,590cc). In addition, the elongated skull, the high forehead and the vault higher than the Neanderthals, the supraorbital protuberances well marked, but not in a burr or torus, the wide face, the narrow nose, appreciable prognathism, low and rectangular orbits, and jaw robust with prominent chin. The tibias very flattened transversely (platycnemia). There was a close relationship between Cro-Magnon Man and Neanderthal Man during the early stages of the Upper Paleolithic in Europe, an area in which there were populations of both species for a brief period - until about 29,000 years ago, or even about 27,000 years in the south of the Iberian Peninsula.

The only certainty is that the Neanderthals became extinct. However, these two human races appeared so suddenly that there are no real indications of their origin. Authors Max H. Flint and Otto O. Binder wrote in their book, Mankind, Child of the Stars: *"Cro-Magnon man appeared with mysteriously enhanced skeletal features, and with a cranial capacity that is staggeringly in excess per 100 cubic centimeters of that of modern man... A similarly large degree of brain expansion occurred in absolutely no other species on Earth in all past ages, nor has any gene shown evidence of brain mutation of comparable magnitude since antiquity."*

The Mysteries of the Grimaldi Caves

At the beginning of the 20th century, scientists began to explore some caves in Italy, very close to the French border. In this process they discovered some burial places as refuges for ancient men, where they apparently resided for a long time. The first remains found show a dating of about 37,000 years. Some of the skeletons found there still wore necklaces and bracelets with which they had been buried. Two of the skeletons found are quite puzzling as the shape of the bones looks like those of Negroid creatures, according to experts. The fact is that no other Negroid skeleton from the same period has been found in Europe.

CHANCELADE'S MAN

It is a human type from the Upper Paleolithic belonging to what scientists call "Homo Sapiens Fosilis", and is associated with the Magdalenian culture. Its vestiges were found for the first time in 1888 in the Dordogne (France). The Chancelade man had an average height of 1.60 m, his skull dolichocephalic, sagittal suture and high forehead. Corpses have been found buried in a shrunken and forced position, smeared with pulverized ocher and with their heads protected by flagstones. The grave goods consisted of necklaces of shells and perforated teeth and bracelets. The ocher and the jewels were something that started from the same order of the Grimaldi bodies. About this skeleton it was discussed if it was the origin of the Eskimos. Currently it is considered a variation of the Cro-Magnon breed.

Homo Floresiensis: the Hobbit Race

As a contrast between the tall men of the Upper Paleolithic we have the "hobbits". At the end of October 2004, a team of scientists from the universities of New England (Australia) and Wollongong (Indonesia) announced in the journal Nature the discovery of a human species that lived until just 12,000 years ago on a remote Indonesian island called Flowers. From the bones found in the Liang Bua cave, it has been concluded that they were men the size of a standing chimpanzee and brains the size of a grapefruit. Despite its limited cranial capacity: 380cc. They knew fire, made stone tools, and hunted in packs. How did its predecessor Homo Erectus come to be isolated in this area? And how is this creature different from the Hobbits and Halflings told in the old novels? Sometimes, not to say ALWAYS, truth is stranger than fiction!

DENISOVAN MAN

DNA from a 40,000-year-old human finger bone found in a Siberian cave points to a new lineage of ancient humans, it is known. The find - the first made with genetic evidence and not fossils - suggests that Central Asia was occupied at the time not only by Neanderthals and Homo sapiens but also by a previously unknown third lineage. *"This is the most exciting discovery yet to come from the field of ancient DNA,"* says Chris Tyler-Smith, a geneticist at the Wellcome Trust Sanger Institute in Hinxton, UK. The work complicates human history once again, similar to the discovery of the controversial Homo floresiensis — also known as "The Hobbit" — which has upended earlier, simpler ways of looking at early human migrations all over the world. the globe according to the evolutionary hypothesis.

If more than four early humans, including the hobbit, lived around 40,000 years ago, " *the amount of [human] biodiversity...was quite remarkable,*" says geneticist Sarah Tishkoff of the University of Pennsylvania. This, without counting the evidence of giant men also found in that period. A team led by archaeologists Michael Shunkov and Anatoli Derevianko, from the Russian Academy of Sciences in Novosibirsk, found a finger bone in 2008 in Denisova Cave in the Altai Mountains of Russia. The cave, which has many archaeological layers spanning 100,000 years, has yielded stone tools from both modern humans and Neanderthals and a small collection of hominid bones too fragmented to identify, according to the researchers.

The finger bone comes from a layer radiocarbon dated to between 48,000 and 30,000 years ago. Evolutionary geneticists Svante Pääbo, Johannes Krause, and colleagues at the Max Planck Institute for Evolutionary Anthropology in Leipzig, Germany, obtained a 30-milligram sample and extracted and sequenced all 16,569 base pairs of its mitochondrial DNA (mtDNA) genome, using new techniques from Pääbo's group that they have successfully used to sequence DNA from both so-called Neanderthals and prehistoric modern humans – let's remember that they are the same thing; cousins. The researchers compared the new mtDNA sequence with that of 54 living people from around the world, one modern human from around 30,000 years ago from elsewhere in Russia, and six Neanderthals.

They got a big surprise: Although Neanderthals differed from modern humans by an average of 202 nucleotide positions in the mitochondrial genome, Denisovans differed from modern humans by an average of 385 positions and 376 from Neanderthals, according to the team reports today in the online edition of Nature. When mtDNA from chimpanzees and bonobos was added to the mix, the researchers were able to estimate that the new hominin shared a common ancestor with Neanderthals and modern humans 1 million years ago. But who was this mysterious human? The study team says the date is too late for an Asian Homo erectus that, according to evolutionary theory, first migrated out of Africa 1.8 million years ago. And in that order, it would also be too early for Homo heidelbergensis, which is believed to have appeared in Africa and Europe around 650,000 years ago.

"There is no evidence" that these or any other known species *" persisted this late "* on the Asian continent, says paleoanthropologist Russell Ciochon of the University of Iowa in Iowa City. Chris Stringer, a paleoanthropologist at the Natural History Museum in London, says the new species could represent a *"pre-heidelbergensis and post-erectus dispersal"* out of Africa *" that we have not observed until now."* For now, Pääbo's team has not given the new lineage a species name, at least until more is known about it. The next thing the researchers plan to do is sequence the nuclear DNA of the finger bone. If they succeed, they could discover the secret identity of hominid X (Michael Balter, March 24, 2010, news.sciencemag.org).

MUNGO MAN OR NEW GUINEA Man

It is thought that around 40,000-50,000 years ago, in the Pleistocene, the first Australians arrived from Southeast Asia. Those first settlers would have traveled from island to island, using the land bridges that linked many of them at that time, and traveling short sea stretches until they reached the eastern end of the Lesser Sunda Islands and the island of New Guinea, to then move by the Australian continental shelf, then above sea level. The reason for such dangerous trips - accepting the theory - is still an enigma. The oldest human remains found to date, Mungo Man, date back 40,000 years, but experts believe that the first human migrations could go back as far as 125,000 years, although this date is disputed.

The remains of Mungo Man were found in New South Wales, some 3,000km off the north coast of Australia where the first human settlements are thought to have been made. Remains consisting of two prominent fossils were found at Lake Mungo: Lake Mungo (also known as Mungo Lady, LM1, or ANU-618) and Lake Mungo (or Mungo Man, Lake Mungo III, or LM3). Lake Mungo is located in New South Wales, Australia, a World Heritage Listed Willandra Lake District. LM1 was discovered in 1969 and is one of the oldest known cremations in the world. LM3, discovered in 1974, was an early human inhabitant of the mainland of Australia, believed to have lived between 68,000-40,000 years ago, during the Pleistocene. The remains are the oldest anatomically modern humans found in Australia to date, although their exact age is a matter of dispute.

The body was sprinkled with red ocher, the same burial system as with the Cro-Magnon men in France. How did two men from the same time and in regions thousands of equidistant kilometers use the same way of burial? This aspect of the discovery has been particularly important to indigenous Australians, as it indicates that certain cultural traditions have existed on the Australian mainland for much longer than previously thought. The new studies show that, using the lengths of his limb bones, the estimate for LM3's height is likely to have been around the abnormal 196 cm (77 inches or 6 feet 5 inches), a rarity for prehistoric men according to history. accepted hypothesis. In 2001, a section of the hypervariable region (HVR1) was published and compared with several other sequences. More than expected number of sequence differences were found compared to CRS humans. This was higher than expected, but did not differ significantly from a mitochondrial DNA insertion on chromosome 11, which is often what is seen in modern human populations. This man did not seem so archaic then.

Fossils of an Unknown Human Species Found in China

An article by Jean-Louis Santini (AFP, March 14, 2012) talks about Stone Age fossils found in China that seem to belong to a hitherto unknown, or at least unclassifiable, human species. Fossils of at least three individuals show that these beings had very different anatomical features, an unusual mix of archaic and modern human features, Chinese and Australian paleoanthropologists noted in the American scientific journal PLoS One (Public Library of Science One). It is the first time that remains of a new species that lived so close to the current period have been found in East Asia, the experts said, adding that these individuals lived between 11,500 and 14,500 years ago.

Until now, one had to go back more than 100,000 years to find, in this part of the world, fossils that were not of Homo sapiens. This species was contemporary with modern humans from the beginning of agriculture in China, one of the oldest in the world, said the researchers, led by Professors Darren Curnoe, from the University of New South Wales (UNSW), in Sydney, Australia, and Ji Xueping of the Yunnan Institute of Archeology (South China). Paleoanthropologists were cautious, however, regarding the classification of these fossils, due to the unusual mosaic of anatomical features they reveal. *"These new fossils could also be of a hitherto unknown species, one that survived until the end of the Ice Age around 11,000 years ago,"* Curnoe said.

The fossilized remains of at least three of these individuals were found in 1989 in Maludong, or 'Red Deer Cave' in Chinese, located near Mengzi, in the Chinese province of Yunnan (south), but they were not studied until 2008. A Chinese geologist discovered the fossil of a fourth partial skeleton in 1979 in another cave near the town of Longlin in the Guangxi Autonomous Region bordering Yunnan. The fossil remains were embedded in a block of rock until 2009, when this same team extracted and reconstituted it. The skulls and teeth from the Maludong and Longlin caves are very similar, revealing a mix of modern and archaic features, features that have never been seen before. Although Asia has more than half the world's population, paleoanthropologists know very little about how modern humans evolved after their ancestors settled in Eurasia 70,000 years ago, Curnoe noted.

Research in this sense did not prosper due to the lack of fossils in Asia and the little knowledge of the importance and antiquity of the remains discovered. "*The discovery of these new humans, dubbed 'red deer people', who hunted for food, opens a new chapter in the story of human evolution - the Asian chapter - and it is a story that is only just beginning to be told.*" Curnoe said.

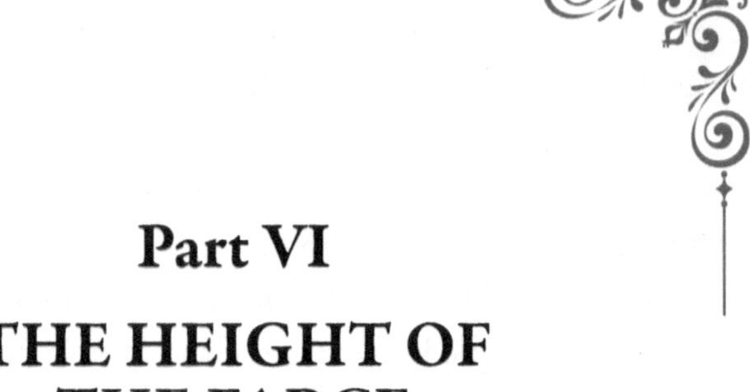

Part VI

THE HEIGHT OF THE FARCE

THE BEST FRAUDS IN *Evolutionary History*

In the fabulous list of imaginary beings, the theory of evolution has added ape-like men of all kinds. Many of them are not just tendentious figures of hairy or stooped men, with thick noses and dark skin, but directly have been created intentionally. Let us remember that a doctor from the Irish army, named French Dubua, found in 1891 a piece of skull and a femur, and that, from these, they made an ape-man, and they called it pithecantropus erectus. But instead of exposing it to the public, they hid it until 1923. It was never released for analysis by the scientific method. When scientists studied them, they ruled that these bones simply belonged to a normal man. As we mentioned previously, their remains supposedly dated back about 50,000 years, but all the evidence has disappeared, so there are no means to ensure anything true about them.

The most interesting thing to note is that the cap of the skull - like that of a monkey - that Dubua found was very far from the femur - like that of a man -, which rules out that both were related. In 1973, MH Day and TI Molleson concluded that "*the gross anatomy, radiological (X-ray) and microscopic anatomy of the Trini femur - where this piece of bone was found - do not significantly distinguish it from the femur of a modern human*." They also said that Homo erectus femurs from China and Africa are anatomically similar to each other, and distinct from Trini's. The fossil record has failed to document a single verifiable "missing link" between ape and man. Compilations of superficial and imprecise evidence, highly speculative constructions, and artist interpretations abound; but there is no scientific evidence documenting the "missing link". "Positive findings" of a "missing link" are periodically announced and subsequently embroiled in controversy, revised, or denied.

Regarding Sinanthropus erectus or Peking Man, despite the disappearance of the remains, it was said that other contemporary skeletons in the area were found. An article about Zhoukoudia appeared in the Scientific American magazine, where Chinese scientists Wu Rukang and Lin Shenglong said they discovered several skeletons where a significant change in the skeleton was seen from 460,000 years ago to 230,000 years ago. What Wu and Lin did not disclose was that the smaller skeleton was not the earlier process before cranial evolution, but rather the remains of a child. So there was no evolution but a tendentious statement that further highlights the theory because then Homo erectus would have coexisted with modern humans, since the secondary ground of the discovery dated back 230,000 years. Preliminary evidence shows that Homo sapiens sapiens, Homo sapiens neanderathalensis, and Homo erectus coexisted in China around the Middle Pleistocene. That is not an evolutionary process.

THE ORCE MAN

The story of the Man from Orce comes from a fossil discovered in 1982 near the Spanish town of Orce, more specifically in Venta Micena. It is designated as VM-0. Gish (1985) reports that on May 14, 1984, French experts revealed that this Hombre de Orce was really a fragment of an ass skull.

Kanjera Man and the Jaw of Kanam

They were discovered by Louis Leakey near Lake Victoria in 1932, they were taken to be very old. But the dating date was very uncertain, since they were probably the bones of modern man. However, modern studies have shown that the morphology of the Kanjera skull is modern, even though the layer in which they were found dates to between 400,000 and 700,000 years old. At another point Leakey found a jaw and thought it was very similar to that of Homo sapiens, although it was estimated to be 2 million years old. Tools and utensils related to prehistoric animals such as Mastodons and Deinotherium were found at both sites.

THE NEBRASKA MAN

With the scientific name of "Hesperopithecus haroldcookii", this "being" was scientifically formed only on the basis of "a tooth" found in 1921 in Nebraska, USA; it was later found to be the tooth of "an extinct pig."

PILTDOWN MAN

Scientifically known as "Eoanthropus dawsoni", this was one of the biggest frauds carried out in the history of evolutionism. It was discovered in England by an amateur, Charles Dawson, between 1908 and 1912. The discovery consisted of parts of an apparent modern skull associated with an ape jaw. Later (between 1913 and 1915) new fragments were found which also appeared to have a mixture of ape and human characteristics and which quelled doubts that the bones came from two creatures. But in 1953 it was discovered to be a hoax consisting of a modern human skull and an orangutan jawbone. It took 37 years for it to become clear that said "Piltdown man", exhibited in the British Museum as the most important evidence of evolutionary theory, was a forgery.

Still, evolutionists continue to develop more sophisticated tricky methods to search for transient non-existent beings. Among the people who were suspected of having carried out such a fraud were different well-established English scientists, such as Sir Walter Elliot Smith or Sir Arthur Smith Wuthword. The story appeared like this: «*In 1912 Charles Dawson, an amateur paleontologist produced some bones, teeth and some primitive tools that he supposedly found in a gravel hole in Piltdown, Sussex, England. In October 1956, Reader's Digest magazine published an article abridged from Popular Science Monthly titled "The Great Piltdown Hoax". A new fluoride absorption method for bone dating revealed that the Piltdown bones were fraudulent; the teeth had been sharpened and, along with the bones, had been discolored with potassium bichromate to hide their true identity. All the "experts" had been deceived for over forty years.*»

Why did it take so long for the scientific community to catch on to the fraud? According to The Scientist of March 15, 2005, Robert Foley, director of the Leverhulme Center for Human Evolutionary Studies, at the University of Cambridge, explained that one of the reasons why The Piltdown Man hoax was so successful "*because it fit the expectations of what people thought early humans would look like.*"

SCHOLAR FRAUD DISPROVES Dating of Neanderthals

Professor Reiner Protsch von Zieten told fellow scientists that a certain 36,000-year-old skull fragment was the link between ancient Neanderthals and modern man. Among his other outstanding discoveries are the remains of a woman who lived 21,300 years ago and those of a man who lived 29,400 years ago. For a long time, specialist dating based on the carbon-14 system had been considered proof that Neanderthals had lived in northern Europe and coexisted, as a separate species, with anatomically modern men. Only there was a problem. The professor did not know how to operate his carbon-14 dating equipment, and the real experts simply concluded that he had made the dates up.

The skeleton that he had dated to be between 21,000 and 36,000 years old was dated by others to be much younger. One of the skulls turned out to be that of a man who had lived just over 250 years earlier, around 1750. On February 19, 2005, an English newspaper reported that the University of Frankfurt had forced this professor to retire, due to his *"many falsehoods and manipulations"* in his 30-year academic career. The scandal erupted when he was caught trying to sell the university's collection of chimpanzee skulls. In addition to fabricating data, another investigation found that he had plagiarized the work of other scientists and passed off fake fossils as authentic.

"This is extremely embarrassing," said Professor Ulrich Brandt, who led the research. *"Of course, the university feels very bad about this."* As a result, Thomas Terberger, a professor at the University of Greifswald in eastern Germany, said: *"Anthropology is going to have to completely revise its concept of modern man between 40,000 and 10,000 years old."* Why did Professor Protsch concoct this fraud? *'If you find a skull that is more than 30,000 years old, that's amazing,'* explained Professor Terberger. «*If you find three of them, you will be*

known to people. That's good for your career. Ultimately, it was out of ambition."

FROM AUSTRALOPITHECUS to Homo sapiens

Many of the evolutionary theories state that we come from a species of algae. Now, where did the eukaryotic cell that makes us as human beings come from these days? That's a big evolutionary problem. However, in the list of fossils there is also an endless fable, starting with Ramapitecus, Australopiecus and Paranthropus.

«Compared to hominids, Darwin says that "an incredible" mutation in human evolution occurred in the last step (in animal evolution these steps do not occur), from very small skulls to larger ones, which are almost transformations into time of evolution, but in the man according to Darwin it happened suddenly. That is, in 300,000-100,000 BC, man had nasals 5 times larger than today, eyebrow bones, and protruding eyes to evade sunlight, but these formidably fast adaptation characteristics do not exist in humans. Observing the skull of the australopithecus and homo sapiens, it is understood that this change must have been a great mutation with the name of "miracle" in a few thousand years, since we assume that a gene mutates every billion years and that miracles do not occur. they are scientists." (Harun Yahya, The Deception of Evolutionism)

Who are our ancestors? Here we only find two groups: monkeys (australopithecus, homo habilis, homo rudolfensis) and men (homo erectus, homo eragster, homo ancestor, Heidelberg man, Neanderthal, Cro-Magnon). What we have here is the same as today: fully human humans, and monkeys entirely monkeys.

KILLINGS AND FRAUDS to Justify a Non-Existent Evolution

After the publication of the first famous book by Charles Darwin, a great campaign began to find the supposed fossils that could be presented as evidence in support of the so-called theory of evolution. Archaeologists began searching for fossils of imaginary creatures called "transient forms." For decades they dug in different parts of the globe, evidently without any success. The disappointment of the case led them to the invention of "Piltdown man" in 1912 and a few other beings out of context. They also tried to assign types of monkeys the quality of semi-humans, and other men presented them as semi-monkeys. Meanwhile, some evolutionists strongly supported the idea of the existence of "living fossils."

According to this belief, if the human race had monkeys as its ancestors, somewhere in the world there should be some half-human beings that have not yet completed the evolutionary process. In this way, towards the end of the year 1800 they found their victims. These were none other than the natives of Tanzania, called "aboriginals", who were designated as "living evidence of evolution." The different structure of the orbit and the relatively heavy lower jaw of the aborigines were the main excuses for defining these human beings as "transitory forms". Evolutionary archaeologists, and many "fossil hunters" who joined them, began digging up the aboriginal graves and brought the skulls to Western evolutionary museums, promptly distributing them to every institution and school in the hemisphere as confirmation of their belief. his evolutionary theory.

"Fossil hunters" did not hesitate to become "headhunters" when the number of graves was not enough to cover their needs. Since the aborigines represented "transient forms" for his famous hypothesis, they had to be considered as animals rather than as human beings. So the "headhunters" murdered the aborigines and legitimized that act by stating that they were doing it in favor of science. The skulls of the hunted natives were sold to museums after subjecting them to some chemical treatments that made them look older. Bullet holes were filled in as carefully as possible. According to "Creation Magazine" published in Australia, a group of observers from South Galler were shocked to see dozens of men, women and children killed by evolutionists.

Among the murdered, 45 skulls were chosen, from which the tissue that covered them was removed, and they were boiled. The top 10 were packed for shipment to England. «*Even today we can see thousands of aboriginal skulls in the depositories of the Smithsonian Institution. Some belong to the corpses extracted from the graves, while others are from innocent people killed to vindicate their theory of evolution. Among the African victims of evolutionary violence, the best known was the pygmy Ota Benga. [...] Since the hypothesis of evolution was not just another scientific theory or hypothesis, but an "ideology" that had to be vindicated, its defenders committed or approved the massacres carried out without the slightest hesitation. To those people, even the massacre seemed legitimate, for the justification of the "lie." That is why such a "lie" is the foundation of the world order that evolutionists erected.*" Harun Yahya affirms in his prestigious book.

A LIE DISGUISED AS *truth*

Harun Yahya rightly wrote: "*Most people are almost completely convinced of the validity of the theory. Therefore, it never asks questions that go to the "how?" and to the "why?" about the theory of evolution while maintaining credibility. In the most "scientific" books about evolution, such is the simplicity with which the state of "transition from water to Earth" is explained—one of the points impossible to explain in the theory of evolution—, that even evolutionists don't believe it. According to the theory of evolution, life began in water, and the first animals to develop on Earth were fish. According to the evolutionary fable, one day some species of fish displayed the ability to get out of the water and move on Earth! The theory goes on to say that the fish that decided to live on Earth had feet instead of fins and lungs instead of gills! In most books on evolution no one explains "why" or "how" the transition occurred. Even in the most scientific sources, writers jump off the bat to conclusions like this: "and the transition from water to Earth occurred," without providing a satisfactory answer as to how the process was accomplished."*

But how did that transformation happen? It is obvious that a fish cannot survive out of water for more than a minute or two. If we accept that there was a dry season, as evolutionists say, and that for some reason the fish were dragged to Earth, what would have happened to them if this process lasted ten million years? The answer is clear: the fish that left the water would inevitably die very soon after. They would all die, one by one. No one would dare say: "It may be that after four million years some of the fish suddenly acquired lungs in order to survive." Surely this would be an illogical statement. Yet that is exactly what evolutionists say. "The transition from water to Earth", "the transition from Earth to air" and millions of other "transitions" are explained by meaningless statements. Evolutionists never mention how complex organs such as eyes, ears, etc. came into existence... »

« Anyway, it is very easy to strongly influence ignorant people by calling these lies "scientific". All one has to do is sketch imaginary representations of "the transition from water to Earth," coin them, and give Latin names to the "grandchildren" living on Earth and to "intermediate creatures," which in most cases cases are imaginary animals. Then comes the time to tell the lie: "Eusthenopteron transformed into Crossopterygian Rhipititistion, and then into Ichthyostega after a long process of evolution." If this is said by a serious-looking "scientist" with black-framed glasses, the statement becomes entirely convincing to most people. The mass media communicate this invention to millions of people around the world, fulfilling their Masonic duty with great excitement. For the majority

of those people who do not have any other opinion than the one given by the media, the "rigorous evidence" they give in this regard becomes highly satisfactory. »

« Another type of lies are the "reconstruction" sketches made by evolutionists. When we look at the publications of evolutionists we come across drawings that represent "family circles", where you see half-human, half-monkey creatures. For example, the fossil displayed as the skull of an Australopithecus (Ardepithecus) is actually the fossil of a Homus erectus discovered at Koobi Fora in 1975. These upright creatures, with hairy bodies and half-human, half-monkey faces, were drawn by the evolutionists who based themselves on the supposedly discovered fossils. However, those drawings are not reliable since a fossil can only provide information about the structure of bones. A scientist can never estimate how much hair was on the body of the fossil. Also, those fossils do not reveal any information about the nose, ears, lips and hair. In their illustrations, however, evolutionists make special point of the half-human, half-monkey noses and lips. This is how evolutionists obtain the drawings of imaginary transitional forms. They even sketch different types of faces for the same type of skull." "Australopithecus Robustus (Zinjanthropus) is a well-known example of such illustrations. It has drawn three types of reconstructions, totally different from one another. [...] the imaginary "Nebraska man", who is drawn next to his "family circle", is an example of how far the imaginative capacity of evolutionists can go.» Adds the Turkish Harun Yahya in his work 'The Deception of Evolutionism'.

INFORMATION SUPPRESSION Methods

The pattern of data suppression has been going on for a long time. In 1880, JD Whitney (a famous California State Geologist), published an extensive article on the advanced stone tools found in California's golden days. The implements, including spearheads and stone mortars, are found deep in mine shafts, beneath thick layers, undisturbed by lava, in formations ranging from 9 million to over 55 million years old. WH Holmes of the Smithsonian Institution, considered one of California's greatest critics, wrote: "*Perhaps if Professor Whitney was fully hesitant to announce the conclusions drawn [that human beings existed in very ancient times in North America], despite the matrix of the imposition of testimony with which he had to deal.*»

In the 1950s, Thomas E. Lee of the National Museum of Canada potentially considered advanced stone tools from glacial deposits at Sheguiandah, on Manitoulin Island in northern Lake Huron. Wayne State University geologist John Sanford said the oldest tools at Sheguiandah are at least 65,000 years old and could be as old as 125,000 years. For those who adhere to standard views on North American prehistory, such ages were unacceptable. Humans supposedly entered North America from Siberia about 12,000 years ago.

Thomas E. Lee complained: « *The discoverer of the site [Lee], was hounded from his Civil Service position to prolonged unemployment; Publication outlets were hacked, evidence was falsified by several prominent authors...; tons of artifacts vanished into storage bins at the National Museum, which had proposed having a monograph published on the site, he was fired and sent into exile; Prestigious official positions and power are exercised in an effort to gain control of just six copies of Sheguiandah that had not gone under cover; and the site has become a tourist destination... Sheguiandah would have been shamefully forced to admit that the Brahminds did not know everything. It would have forced the rewriting of almost every book in the business. It had to be killed. He was murdered.*"

Part VII

THOSE MEN BEFORE ADAM

"The clergyman turned the simple teachings of Jesus into an engine to enslave humanity, and adulterated by artificial constructions into an invention to steal wealth and power for themselves... this clergy, in fact, constitutes the true Antichrist."
Thomas Jefferson (Former US President)

WHEN DID MODERN MAN Appear?

Paleoanthropologists believe that anatomically modern humans (Homo sapiens sapiens) arose piecemeal from Homo erectus. Around 300,000 or 400,000 years ago. Homo erectus are described as having a cranial capacity nearly as large as that of modern humans, yet some of its characteristics, such as the thickness of the skull and a large, ridged, receding forehead, still manifest to a lesser degree. Examples in this category are the finds at Swanscombe in England, Steinheim in Germany, and Fontechevade and Arago in France. Because these skulls also possess, to some extent, the characteristics of Neanderthals, they are also classified as pre-Neanderthal types. Most authorities now postulate that anatomically modern humans and western European Neanderthals evolved from pre-sapiens Homo Neanderthals or early hominin types less than a million years ago.

It is important to clarify this point before proceeding. In the early part of the 20th century, some scientists support the idea that the Neanderthals of the last ice age, known as the Classical Neanderthals of Western Europe, were the direct ancestors of modern humans. They had a larger brain than Homo sapiens sapiens. Their faces and jaws were much larger, and their foreheads were smaller, receding back behind large arches of the eyebrows. The remaining Neanderthals are found in Pleistocene deposits ranging from 30,000 to 150,000 years old. However, the discovery of the earliest Homo sapiens in deposits much older than those 150,000 years effectively removes the classical Neanderthals of Western Europe from the direct line of descent from Homo erectus to modern humans, thereby the link between the two is lost.

Furthermore, the type of humans known as Cro-Magnon who are believed to have appeared in Europe a little over 30,000 years ago, were anatomically modern. One scientist used to say that the anatomically modern Homo sapiens sapiens first appeared about 40,000 years ago, but now many authorities, in light of discoveries in southern Africa and elsewhere, say it appeared 100,000 years ago or more. They do not agree, not even the evidence supports their hypotheses. As we can see, all these skeletons of apes and men, including the imaginary ape-men, are no more than 2 million years old, the time at which the fantastic and incredibly fast human evolution from primates to modern man would begin, according to the paradigm accepted.

But, what would happen if we knew that archaeologists have found solid evidence of non-archaic human beings or any kind of caveman with a chronological estimate greater than those 4 million years? If this were so, even the genesis narrative of Adam and Eve would be false or perhaps misunderstood. All this would lead us headfirst into new theories about the prehistory of our globe and of life.

In Search of the Oldest Woman

Studies by the American geneticist Douglas C. Wallace, from Emory University, Atlanta, led him to come up with an answer going back 100,000 years ago: « *There was a woman who possessed the DNA of these mitochondria [studied]. If it was the only one, it was Eve!*» Therefore, Homo erectus dated between 1.6 and 2.5 million years ago had nothing to do with "our Eve". A group of anthropological researchers from the University of Berkeley, California, came to the conclusion that the gradual dispersal of man came from Africa about 180,000 years ago. This again demolishes the claim that Java or Peking man were human ancestors. For their part, the geneticists JS Jones and S. Rouhani from University College London, and a group headed by Jim S. Wainscoat, from the University of Oxford, studied the geographical distribution of betaglobin among 8 population groups.

The answer was that there was a founding population in Africa that was threatened with extinction. That is, in a nutshell, that in the event that the founding population can be reduced to a maximum of three couples or even a single female, then the millions of years of old bones and skeletons only prove that all the bones and skeletons that presented to us in this absurd theater do NOT belong to our direct ancestors, but represent the relics of our "indirect" ancestors; As is understandable, they could not refer to the "founding population" either, since it is believed that this did not appear until about 180 or 100 thousand years ago. Evolutionists say that if there was a common origin it must come from Africa about 180,000 years ago, however, if this were so, there should be no evidence of humans older than this date.

Doctors Willibald Katzinger, director of the Nordic Museum in Linz, Austria, and Hans-Joachim Zillmer visited Professor Gutiérrez Lega in Bogotá, Colombia, where he showed them evidence of prehistoric humans. In his collection, Gutiérrez Lega has a human hand and foot covered with lydite, as could be scientifically determined in Vienna. The sedimentary stone (lydite) belongs to the Mesozoic Era, the age of the dinosaurs, and was found in Colombia at an altitude of 2000 m, along with other fossil remains belonging to the same period. So, who was this original race from which the scientists concluded that the rest of the humans would have come? Before concluding something, we must know that the Sumerian culture says that man was created genetically about 270,000 years ago and left in South Africa. Could this have some kind of relationship? At least the archaeological evidence does not say that man is so "recent."

THE TRUE RECORD

"The human race is not native to this planet." (Michael A. Cremo. Writer of Hidden History of the Human Race) Given the massive amount of evidence on findings that humans have existed for as long as the geological history of planet Earth itself, we will try to be brief and cite the most significant cases. As we mentioned earlier, evolutionary history assumes that there can be no finds related to modern humans older than 250,000 years. If everything unearthed from the bowels of the Earth were brought to light, we could refer to so many cases of human appearances from 250,000 years ago - including very advanced technology - to write an encyclopedia, but we will refer to the oldest of the that knowledge has been given.

The evidence seriously suggests that humanity already lived in prehistory, and with more technology than we have today, however, by using it in the wrong way, they were forced to live in caves and lose all knowledge and social relationships that they had. they had Much of this information can be found in Richard L. Thompson and Michael A. Cremo's book: "Forbbiden Archeology". «*In the last 50 years archaeologists and paleontologists have buried as much evidence as they have unearthed, systematically.*» (Richard L. Thompson. Writer of the book Forbidden Archaeology)

In the Paleozoic era (544 million years to 245 million years) the Earth did not flourish in an evolutionary process, rather life arose abruptly, and we see this when examining the terrestrial strata and fossil records. It becomes clear that all living organisms appeared simultaneously. One of the oldest terrestrial strata where fossils of creatures from another era were found is the Cambrian, with an estimated age of 500-550 million years. According to the fossil records, the creatures found in the levels of that period all appeared suddenly, that is, without ancestors that preceded them. Fossils found in Cambrian rocks belong to snails, trilobites, sponges, earthworms, jellyfish, sea urchins, and other complex vertebrates. This vast mosaic of living organisms comprises a large number of complex creatures which, appearing so suddenly as a truly miraculous event, was dubbed the "Cambrian Explosion" in the geological literature. Must take into account:

1. The mammoth was extinguished by indiscriminate hunting by man in relatively recent times.
2. Many species of prehistoric crocodiles were part of the Neanderthal diet.
3. The saber-toothed tiger was a species that became extinct and resurfaced like the phoenix, many times.
4. The coelacanth is still alive, it was fished in 2009, it is believed to be the oldest fish in existence.

...AND BEFORE THE PRIMATES Man Existed

To begin this section we have to refer to the caveman, who should not have existed 2 million years ago, however, they have been found:

1. Prehistoric shark teeth with holes similar to those made by many tribes and cultures to make chains or amulets. A significant case is the one discovered in the Red Crag formation in England, indicating an age of 2.0 to 2.5 million years, found by Edward Charlesworth (Member of the Geological Society) who presented it at the Royal Institute of Anthropology. of Great Britain on April 8, 1872.

2. A shell in which a human face was represented. This shell was exposed by H. Stopes (Fellow of the Geological Society) in 1881. The shell also belonged to the Red Crag formation and is between 2.0 and 2.5 million years old. This was also quoted in The Geological Magazine in 1912 when it was denied that it was a forgery.

3. A modern human jaw was found in Foxhall, England, and along with it evidence of the use of fire and cutting tools. This mandible was dated to about 2.0 to 2.5 million years ago. J. Reid Moir highlighted its importance in 1927 as a Fellow of the Royal Institute of Anthropology and President of the East Anglian Prehistoric Society. It was found by excavation workers in 1855; Another mandible, in this case with two molars and with the same age, was found in Miramar, Argentina, and reported by MA Vignati in 1921, although it was officially found by Lorenzo Parodi, a museum collector. A few years earlier, the discovery of other eolithic tools and land burned by human action had already been reported, also in Miramar.

4. Eolithic evidence (primitive use of quartz in its original state to make tools) in the Soan Valley, Pakistan, by British archaeologists. These tools are 2 million years old. In Monte Aperto (Italy) spearheads and bones with incisions made by human action were also found. This was reported by the

professor of geology at the University of Bologna, Mr. G. Capellini, on November 25, 1875.

5. Clay human figures discovered in Nampa, Idaho, in 1889. These are also 2 million years old (Pliocene and Pleistocene period, between 7 million and 400,000 years). The figures were found in a mine shaft 100m deep in a sedimented clayey layer. It was less than 4cm tall and represents a female figure as perfect as the best sculptures of Classical Greece.

6. Bones of extinct animals with incisions from human bones were found in the Arno River, by J. Desnoyers, and in San Giovanni, Italy, by Professor Ramorino. These secondaries were exhibited at the Italian Society of Natural Sciences in Spezzia on September 20, 1865. This is just an indication of these many discoveries that are accompanied by animal bones that appear cut as if by the action of human implements or sectioned to remove the bones. organs. This is seen in the case of A. Laussedat who informed the French Academy of Sciences that P. Bertrand had sent him two fragments of prehistoric rhinoceros jaws with these marks, coming from a site near Billy, France. These animal remains are between 3 and 15 million years old – we will cite this case later. Another case of the same magnitude was presented by M. A. Ferretti at a meeting of the Geological Committee of Italy in 1876.

7. The discovery of arrowheads and spears millions - not thousands but millions - of years old are a traditional dish in archaeology. Some interesting cases are, for example, the 2-million-year-old discovered in Janicule (Italy) presented by Professor G. Ponzi in 1871 at a meeting in Bologna of the International Congress of Prehistoric Anthropology. The professor himself also studied other eolithic tools found in Acquatraversa and with the same antiquity.

Human Evidence Out of Time

Human footprints. In the Dinosaur Valley National Park (Texas - USA), in the bed of the Paluxy River and on the rocky plateau on the shore, innumerable petrified human footprints have been found. The most surprising thing was that the tracks were located in much higher geological layers. In principle, the river water should have quickly eroded the tracks left by prehistoric animals until they disappeared, but in reality, tracks that were at least 60 million years old can be perfectly observed. Who were those human beings? If the dinosaurs became extinct 65 million years ago and our humanity supposedly only existed for 3 million years, it is impossible for there to be graphic representations of prehistoric animals. If today's idea of evolution is correct, no man could have ever seen a dinosaur and therefore could never have drawn it in a cave.

In 1908, near Glenn Rose (Texas), giant human hand and foot prints – which would have been 4 m high - were discovered, mixed with dinosaur footprints (between 120 and 130 million years old). At Laetoli, in northern Tanzania, just 30 miles south of the large Olduvai Gorge site, expedition members of the popular Leakey family - bone hunters - led by Mary Leakey reported finding marks on the Earth, some of which were animal footprints and others of a modern human. The footprints were imprinted in layers of volcanic ash, which yielding to potassium-argon dating gave 3.6 to 3.8 million years.

MH Day studied the footprints using photogrammatic methods - which consist in the study of obtaining exact sizes through the use of photography - reaching the following conclusion: «*close similarities with the anatomy of the modern man's foot, habitually barefoot; debatably the normal human condition.*" Concluding: "*There is now no serious dispute as to the upright posture and bipedal gait of australopithecines.*" Finally RH Tuttle maintained: «*The shape of the footprints are indisputably like those common strides of barefoot humans.*»

There is a petrified footprint of a human who walked upright between 5 and 15 million years ago which was found in the current Bolivian altiplano. It is much older than the one discovered in the Siwa oasis, Egypt in August 2007, which would be 2 million years ago. These are remains belonging to the Miocene of the Tertiary epoch, when it is assumed that the Andes mountain range was in formation. This makes it the "certainly oldest human footprint ever discovered." In addition to the shape of the fingers, it could be concluded that this man walked erect. Before this event, a group of Egyptian archaeologists found what they say could be the oldest human footprint in history in the desert of western Egypt.

The footprint was discovered while exploring a prehistoric area in Siwa (part of a desert oasis). The footprint was imprinted on petrified mud in the form of rock. Hence the explanation that it has been able to remain intact to this day. As usual in these cases, the studies have been carried out based on carbon 14 and, given the age of the rock, it is assumed that it could be older than the famous 3-million-year-old fossil of Lucy, the partial skeleton of an ape Found in Ethiopia in 1974.

According to the hypothesis of evolution, the first hominid that walked upright was Australopithecus anamensis, of which remains were found in Kenya, which are 4 million years old. The Bolivian human footprint was found on the shores of Lake Titicaca, almost

70 km west of La Paz between the towns of Tiwanak and Guaqui. The scientists reported that the remains were dated using the same techniques as in the case of the Siwa footprint. This would further complicate evolutionary justifications since it is believed that the first men did not migrate to America from Siberia until only 300,000 years ago.

There is overwhelming evidence that man is older than the hominids presented in the geological evolutionary column, that his ancestor is neither the Neanderthal nor the fossils of the so-called pre-humans, but man with 46 chromosomes - let's remember that hominids including Neanderthals have been on Earth for hundreds of millions of years. One of these pieces of evidence is the balls of unknown alloy from Klerksdorp in South Africa, with an age of more than 2.8 billion years that we will mention later; another more evident is the anatomically foot of a modern human found in Islamabad, in Margalla Hills in July 2007, with more than 1 million years, which proves that man was already man before many of these primates and hominids including the Neanderthal, appeared on the scene.

On October 8, 1922, the American Weekly, section of the New York Sunday American, ran with a prominent headline: "*Mystery of the 5,000,000-Year-Old Petrified 'Shoe Plant,* '" by Dr. WH Ballou. Ballou wrote: "*Some time ago, while he was prospecting for fossils in Nevada, John T. Reid, a distinguished mining engineer and geologist, stopped suddenly and looked down in utter bewilderment and surprise at a rock near his feet. There was, in a part of the same rock, what seemed to be a human footprint! Closer inspections showed that it was not a bare foot mark, but was, apparently, a shoe sole that had turned to stone. The front was gone. But there was at least the outer line of two-thirds of it, and around this outer line ran a well-defined stitched thread that had, it seemed, added the welt to the sole...*" Reid brought the specimen to New York, where tried to bring it to the attention of scientists.

Upon his arrival he introduced him to Dr. James F. Kemp, a geologist at Columbia University; Professor HF Osborn, WD Matthew and EO Hovey of the American Museum of Natural History. They all concluded that it was indeed the best imitation of an artificial object that they had ever seen. The verdict of the scientific class did not satisfy Reid, who commissioned new analyzes and photographs from a chemist at the Rockefeller Institute. The new contributions left little room for doubt: the sole was human work. But who made shoes 200 million years ago? Thus it was finalized that the genuine fossil dating was of rock belonging to the Triassic period, that is, it is estimated at an age of 213-248 million years.

Another particular case is that of the so-called "Guadalupe Man": «*W. Cooper exposed in 1983 that a modern skeleton had been found during the year 1812 in Guadalupe (Mexico) with an age of 25 million years.*» But the fact that the skeleton was next to a dog and some instruments indicates that the dogs themselves would not have evolved, and those humans would already be workers with prefabricated tools in those times. These skeletons are human remains found on an island in the Antilles, but with the peculiarity that they were found in a stratum with a geological dating of at least 28 million years, that is to say, from the Miocene epoch, long before the beings modern humans to appear on the island. For many researchers the dating is not correct, but the debate is still open.

One of the samples taken from the coast of Guadeloupe, near the village of Moule, was a stone slab weighing about 2 tons that was sent to the British Museum in 1812, where it was exposed to the public, but with the arrival of the Darwin's theory, the slab was relegated to the basement. One of the things in favor is that these remains have been studied scientifically and can still be seen today in the

British Museum. This is reminiscent of a report from 1983, where the Moscow News gave a brief - but intriguing - report in which a human footprint in 150-million-year-old Jurassic rock appeared next to the giant footprints of a three-toed dinosaur. The discovery was made in the Turkmen Republic, which at the time was located in the southeast of the USSR.

Much has been heard about the nickname "Meister's Man", because in 1968, William J. Meister - a collector of trilobites - discovered a stone that had the outline of a sandal, shoe or boot marked, and under the footprint was a trilobite. It is important to note that trilobites are believed to have disappeared more than 280 million years ago. The discovery occurred in a slate field near Antelope Spring (Utah, USA). The estimated date for its "printing" ranges from 590 to 505 million years. The appearance of this footprint with a boot and its relationship with the trilobite break all established patterns. How is it possible that men with boots walked across North America more than 510 million years ago? The article was cited in Creation Research Society Quarterly.

Meister added the following important piece of additional information: *"On July 4, I accompanied Dr. Clarence Coombs, Columbia Union College, Tacoma, Maryland, and Maurice Carlisle, graduate geologist, University of Colorado at Boulder, to the site of the discovery. After a couple of hours of digging, Mr. Carlisle found a protruding slab, which he said convinced him that the discovery of footprints at the site was a distinct possibility, since this discovery showed that the formation was at one time in the surface.»*

In Spain, they found traces of a type of herbivorous dinosaur that was between 4 and 5 m long, walking at a speed of 5 km/h together with its young and behind them, traces of humans, probably hunters; Also in South America, specifically in Argentina, dinosaur remains were found, with marks on some bones, as if they had been attacked with stone-tipped spears, along with human footprints. Also found in caverns discovered in Peru were the bones of carnivores that supposedly became extinct millions of years before the appearance of humans, as if they had cooked it, split the bone in half and eaten its interior. Weapons made from dinosaur bones and cave paintings that show how a group of 10 men hunted a mammoth; In the Altamira cave, cave paintings show hunters with a certain type of giant, now extinct elk.

Human bones in impossible ages. In 1965, Bryan Patterson and WW Howells made a surprising discovery of a humerus resembling that of a modern human in Kanapoi, Kenya. In 1977, French workers found a similar humerus in Gombore, Ethiopia. The deposit to which the Kanapoi humerus belonged was about 4.5 million years old. A 1975 study by medical anthropologist CE Oxnard concluded: *"We can clearly confirm that the Kanapoi fossil is very human-like.»* An equally significant case occurred in India, where a group of archaeologists found the skull of a woman belonging to the Jurassic Era (about 65 million years ago).

Traveling further into the past we discover that technology increases, instead of decreasing. Thirty miles west of the Italian city of Genoa is Savona, where a modern human skeleton was found in the 1850s while rebuilding a church was underway. This human skeleton was 3-4 million years old. On this, Arthur Issel communicated in 1867 the details to the members of the International Academy of Prehistoric Anthropology in Paris.

In the Italian Alps, several human remains from 3-4 million years ago were found in 1860. Its discoverer was Professor Giuseppe Ragazzoni, a geologist at the Technical Institute of Brescia. Ragazzoni sent the information to several experts who did not want to give credit to his discovery, however, in 1875, Carlo Germani, one of the people whom Ragazzoni notified about his discovery, went to Castenedolo to study the area and ended up finding him. also remains of 3-4 million years and supporting Ragazzoni. This site is rich in all kinds of vestiges and has several prehistoric layers, since apparently it was a part that was formerly covered by water. In this sense, Gabriel de Mortillet reported that M. Quiquerez discovered a modern human skeleton in Delémont, Switzerland, in a layer of ferrous clay from the Late Eocene (38-45 million years). Mortillet also cited the discovery of another skeleton of this category by Garrigou in the Miocene of Midi de France. This means that it would be between 5 and 25 million years old.

This information from the humanities earlier than we know is consistent with findings from Argentina in the 19th century, when paleontologist Florentino Ameghino claimed to have found human remains in tertiary terrain. At the time, these discoveries were viewed with skepticism by the scientific establishment. The same lack of interest deserved a more recent finding by the anthropologist Hernao Marín in Colombia: the fossilized remains of an antediluvian animal (an Iguanodon) appeared mysteriously associated with a Neanderthal man. Among the most incredible discoveries of this kind is one by Richard Leakey who found a human skull under a layer of rock dating back 212 million years.

In 1887, a famous geology and fossil researcher from the Argentine provincial coasts, Florentino Ameghino, found important discoveries in Monte Hermoso, 37 miles northeast of Bahía Blanca. F. Ameghino wrote about it: « *The presence of man, or rather his precursor, in this ancient site, is demonstrated by the presence of crudely* (roughly) *worked flints like those of the Miocene in Portugal, carved bones, burned bones, and burnt earth coming from old chimneys* (hearths).» The age of these deposits is around 3.5 million years. It is recognized that it is not believed that there were even ancestors of man in America millions of years ago.

Florentino Ameghino also found stone tools along with cut bones and signs of fire in the Santacrucia and Entrerrea formations. The Santacrucio formations belong to the Early and Middle

Miocene ages, making the tools found a relic of 15-25 million years. Florentino Ameghino's brother "Carlos" also found great finds in Miramar between 1912 and 1914. One of his greatest discoveries about humans in prehistory came from the Chapamalalan layers where Carlos Ameghino extracted a toxodon femur (a prehistoric mammal extinct) along with other important artifacts, but what was impressive was the discovery that the toxodon femur contained the tip of a stone projectile. All these evidences were around 5-12 million years old.

Another case, this time in California, deals with the doctor HH Boyce, who found human bones 5 million years old in 1853. Other human bones as old as 8.7 million years have been found in many other locations in California. A modern human skeleton in Switzerland, specifically in Delèmont, was found in a layer of the Late Eocene (38-45 million years). On April 13, 1868, A. Laussedat informed the French Academy of Sciences that P. Bertrand had sent him two fragments of a rhinoceros lower jaw. These fragments came from a hole near Billy, France. *«One of the fragments had four very deep grooves in it. These short grooves, located at the bottom of the bone, were roughly parallel. According to Laussedat, the cut marks appeared in cross section like those made by an ax on a piece of hardwood. So he thought the marks were made in the same way, which is to say, with a pointed stone instrument with a handle, where the bone was fresh [...] Just as remote is shown in the facts that the jawbone was found in a formation of the Middle Miocene, about 15 million years ago.»*

Of the same rank are the bones of Halitherium cut by men, cited in the book Hidden History of the Human Race by Michael A. Cremo and Richard L. Thompson on page 12. Another case also occurred in April 1868, coming from the French Academy of Sciences, which contained a report by F. Garrigou and H. Filhol: "*We now have sufficient evidence to allow us to assume that the contemporaneity of Miocene humans and mammals is proven.*" This evidence was a collection of mammalian bones, apparently intentionally broken, that came from Sansan, France. Especially the broken bones found were from Dicrocerus elegans, a small prehistoric deer. Other similar discoveries were made at Pouancé and Clermont, both in France. The relationship between all of them is that they were located in an antiquity between 12 and 19 million years.

Another case corresponds to August 19, 1867, in Paris, where L. Bourgeois presented to the International Congress of Prehistoric Anthropology and Archeology a report on a flint implement that he found in Early Miocene layers (15-20 million years) in Thenay, north of central France. Gabriel de Mortillet was one of the first to take an interest in these finds and also investigated the area and the artifacts. At a depth of 14 feet, belonging to the Early Miocene, Bourgeois discovered many flint tools also in 1869. About this Mr. de Mortillet wrote in Le Préhistorique: « *There were no more doubts about their antiquity or about their geological position.* » Much later, a human skeleton was unearthed in Illinois, United States, in December 1962.

The American magazine 'The Geologist' announced the discovery of a series of human bones found in a coal sediment in Macoupin County, Illinois, (USA). The bones were covered with a very hard shiny crust, as black as the charcoal itself and with the appearance of slate, which after being removed left the bone remains uncovered, presenting its natural state. C. Brian Trask of the Illinois Geological Institute dated the coal to 286 million years old, though it could be as old as 320 million years. A similar event was reported by Professor WG Burroughs, head of the geology department at Berea College in Berea, Kentucky in 1938. What he discovered was clear evidence of modern humans walking on two legs in the High Carboniferous Rockcastle Country. In Pennsylvania this date began 320 million years ago.

Loose limbs. In 1842, a poorly preserved human skull was found in a lignite that was between 15 and 50 million years old. The object is part of the collection of the Freiberg Mining Academy in Germany. Another skull, in better condition, was discovered in India, but this one was 65 million years old. Not so old, but in any case interesting, was the discovery of a human foot in Islamabad in July 2007. It is a modern human foot that was found in Margalla Hills, and whose age is around 1 million years. Another striking case was a fossilized finger, identified as DM93-083, which was found in the Canadian Arctic. It is dated from 100 to 110 million years ago, a time that corresponds to the Cretaceous.

Human Artifacts Out of Time

A piece of gum from the past. A 5,000-year-old chewing gum was discovered in 2007 by a 23-year-old British Archeology student in Finland, Sarah Pickin. He found the age-old chewing gum, made from birch bark tar, in excavations on the west coast of the Scandinavian country. The chewing gum -one of the oldest that have been found- presented its corresponding teeth marks.

Prehistoric flutes. Once again, the German site of Hohle Fels, gives us a new record with the oldest artifact found, where eight flutes between 30,000 and 40,000 years old were discovered. The most spectacular is a 40,000-year-old flute made from a griffon vulture bone that is being presented today in the scientific journal Nature, making it the oldest musical instrument discovered to date. Although one end of the flute is broken, "*the fragment that has been able to be recovered is 21.8 cm long and has five finger holes that allowed complex melodies to be played,*" reported Nicholas Conard, an archaeologist at the University of Tubingen (Germany), and first author of the research, in a telephone interview. Who taught you to make and play harps? Of the eight flutes discovered at three sites in the Swabian region of southern Germany, four are made from mammoth tusks and the other four from bird bones (one from a griffon vulture, one from a swan and two from non-native species) identified).

Half of these eight flutes have turned up in the last excavation campaign, while the other half have turned up in excavations carried out in previous years. The griffon vulture flute, discovered last September 17 at the Hohle Fels site, is a surprising jewel of craftsmanship in those Homo sapiens before cave paintings, agriculture and writing. Not only are the five finger holes carefully carved with stone flakes striking, but also the tiny line-shaped incisions that can be seen next to the holes. These incisions, the researchers believe, were made to indicate the point where the holes had to be carved so that the flute would sound in tune. "*We've built a replica with another griffon vulture bone, carving the holes at the same points, and it sounds the same notes we hear in today's music,*" Conard explains. « *Any tune can be played on this flute* », but the timbre of the instrument is peculiar. "*We tried to play the American anthem and it was reminiscent of Jimi Hendrix's version at Woodstock.*"

The researchers also highlight the technical difficulty of making flutes from mammoth tusks, something "*much more difficult than making them from bird bones,*" Conard notes. You have to carve the tusk fragment with which the flute will be made, cut it in two lengthwise like the bread of a sandwich, empty the two halves inside, carve the holes and put the two halves back together so that they are sealed. The flutes found in Germany "*do not represent the origin of the music*", warns Conard. Before these flutes, he proposes, there must have been other more elementary ones. And before the singing should have existed. And before, even, the rhythm. The archaeologist Eudald Carbonell, co-director of the Atapuerca excavations and a great music lover, highlights: «*Possibly the first human species that lived in Africa more than two million years ago already experimented with rhythm. But making and playing these flutes is something much more advanced that requires a more sophisticated brain. The percussion is ancient, the melody is modern.*"

In the last excavation campaign, along with the flutes, bones of reindeer, bears, horses, mammoths and mountain goats have been found, species with which humans lived in Europe at that time. No human remains have been found, so it is unknown who made them. Some of the oldest sculptures in the world have also been discovered at the same Hohle Fels site where the flutes appeared. Carved from mammoth ivory, one of the figures depicted is a half-man half-lion hybrid. It is dated to 30,000 years old and, when discovered in 2003. More recently, a 35,000-year-old female figure has appeared on the Hohle Fels which has been heralded as the oldest figurative art form discovered to date. The so-called "Venus of Hohle Fels", featured in the May 14 issue of Nature, stands out for having grotesquely sized breasts and vulva, while other features of her body are not exaggerated.

Another figure discovered is that of a sculpture also carved in mammoth ivory, representing an aquatic bird. This is 30,000 years old. Waterfowl were part of the fauna with which humans who arrived in southern Germany through the Danube valley lived around 40,000 years ago, according to accepted theory. The large mammals that Paleolithic humans hunted occupy a prominent place in artistic representations such as the cave paintings of Lascaux or Altamira. Another sculpture, in this case small and representing a horse's head, has also been found in Hohle Fels, as well as remains of other large prey -although not sculptures- such as bears, reindeer and mammoths.

The Stainless Column. In the courtyard of a temple in Delhi, India, there is a column made of welded pieces of iron that have been exposed to wear and tear for more than 4,000 years without ever showing a trace of rust, since they do not contain sulfur or phosphorus.

The Greek marrow seekers. At a site called Pikermi, near Marathon in Greece, there is a rich layer of Late Miocene fossils, explored and described by the prominent French scientist Albert Gaundry. During the 1872 meeting in Brussels of the International Congress of Prehistoric Anthropology and Archeology, Baron von Dücker reported broken bones from Pikermi proving human existence in the Miocene. In this place, 34 parts of the jaws of Hipparion (an extinct horse with three fingers), antelopes and also 19 fragments of tibia and another 22 fragments of bones of large mammals such as rhinos were found. All showed traces of methodical fracture for the purpose of extracting marrow. According to Dücker, all the holes made were *"more or less distinct blast tracks from hard objects."* These layers suggest that the findings range from 5-12 million years.

Carved bone from the Dardanians in Turkey. In 1874, Frank Calvert found in a Miocene formation in Turkey (along the Dardanus) a Deinotherium bone with carved figures of animals on it. Calvert noted: *"I have found in different parts of the same cliff, not far from the site of the carved bone, a flake of flint and some animal bones, fractured longitudinally, obviously by the hand of a man for the purpose of extracting marrow, according to to the practice of all primitive races."* The dating of all this evidence points to an antiquity of around 5-25 million years.

Human jobs in Belgium. In February and March 1918, Wilhelm Freudenberg, a geologist attached to the German army, was conducting tests for military tests on tertiary formations west of Antwerp in Belgium. In clay pits at Hol, near St. Gillis, and in other areas, Freudenberg found carved objects along with bones and cut shells. Most of the objects came from Early Pliocene and Late Miocene deposits, that is, they were about 4-7 million years old, and were obviously human fabrications. Also in Belgium, A. Rutot, curator at the Royal Museum of Natural History in Brussels, made a series of discoveries dealing with anomalous stone tool industries taking on new relevance during the early 20th century.

Much of the industry identified by Rutot dates to the Early Pleistocene. But in 1907, Rutot's advanced research turned up in the Boncelles area, in the Arden region of Belgium. The layers where the artifacts were found were from the Oligocene, which implies that they were between 25 and 38 million years old. Other notable discoveries on the use of human utensils in Belgian prehistory were made by Rutot in Bay Monet and Baraque Michel offering similar results.

Diamonds, spokes and spark plugs. Although among these "OOParts" (Out Of Place Artifacts) we have a diamond that was found in Herkimer, New York. This diamond was inexplicably found "already cut", which leaves a lot to think about since diamond polishing requires specialized, high-precision machines. Among these artifacts we also have an aluminum belt buckle discovered in China from 265 BC, and a Chinese radio from Galena from 2,500 years ago. Another gadget is the Sparkplug spark plug, which is thousands of years old. Who needed spark plugs in prehistoric times? Who had diamond grinding machines? And how and why did they build excavating machines?

Aluminum block. Found in 1973, the artifact was discovered by a group of workers carrying out an excavation on the banks of the Mures River, two kilometers east of the city of Aiud, Transylvania. Three objects were found simultaneously at the same site, of which two were fossil bones belonging to a Mastodon. The third object, the aluminum block, was also housed in stratum number 35 and was clearly different from any common piece of animal bone or geological object. The curious block was donated to the Transylvanian Museum of History, to be rediscovered and analyzed many years later. Its weight turned out to be 5 pounds, and its approximate measurements of 20 x 12.5 x 7 centimeters.

Chemical tests carried out in a laboratory in Lausanne, Switzerland, to determine its composition, showed that the artifact was made up mostly of aluminum (89%), with the minor participation of 11 other metals in specific proportions. The surprise for the scientists was no less, since aluminum in its pure state is not present in nature, and the technology to achieve a considerable degree of purity could only be achieved in the mid-19th century. Another of the millennial inexplicable artifacts is a metal cube discovered in Australia and whose origin is totally unknown.

Modern humans used fire to make tools. Evidence reported in the Aug. 14 issue of the journal Science shows that early humans living on the shores of Africa 72,000 years ago used complex heat treatment to make double-sided blades and tools. Unheated silcrete can show drastic color and texture changes upon heating and flaking. An international team, including three researchers from Arizona State University's Institute of Human Origins, point out that silcrete is found no closer than 5 km from their excavations at Pinnacle Point, Mossel Bay in South Africa, and that most of the pieces found are extremely desquamated.

The discovery places complex cognition at 72,000 years ago, and perhaps much earlier. Evidence has been reported that modern humans living on the southernmost coasts of Africa 72,000 years ago employed pyrotechnology – the controlled use of fire – to increase the quality and efficiency of their stone tool making process, such as and as told in the August 14 issue of Science magazine. An international team of researchers, including three from the Arizona State University (ASU) Institute of Human Origins, conclude that "*this technology required a novel association between fire, its heat, and a structural change in the stone with the consequent benefit of the scales.*" Furthermore, their findings open up the idea of complex cognition in these early engineers. Marean is a paleoanthropologist at the Institute of Human Origins and a professor in the School of Human Evolution and Social Change in the College of Liberal Arts and Sciences at Arizona State University.

The heat transformed a stone called silcrete, which was bad enough for making tools, into an amazing raw material that allowed modern humans to make highly advanced tools. Who taught you this art? Overnight, cavemen went from living in caves, hunting with spears, and wearing rags to learning about agriculture, makeup, warfare, weapon making, astrology, astronomy, magical arts, crafting. of cyclopean monuments, medicine, navigation, languages and alphabets, martial arts, plastic arts, mathematics and science. These things do not happen by chance but because of the teachings of more advanced civilizations and thousands of years of hard progress helped by "masters" who know these arts. These people advanced without internet, telecommunications, television, radio, telephone, press, news, any apparent type of writing, since this would only lead them to increasingly forget a previous knowledge due to the action of the years, the distances, the wars and the death of the primeval knowers.

The tools of Aix-en-Provence. However, other older pieces and tools have been discovered, such as those found between 1786 and 1788, near Aix-en-Provence. Several finds were made in a limestone quarry, where rock layers alternate with layers of sand and clay. About 15 m below ground level, in a layer of sand, some workers first found pieces of columns and carved blocks; then lower down, pieces of metal resembling coins, petrified wooden tool handles, and a large wooden table also petrified. The set would be 300 million years old, if the classical theories of geology are accepted, regarding the formation of the rocks and the period of petrification. This evidence, especially, refutes the evolutionary hypothesis and demerits that homo habilis was the first tool worker.

Weapons. Sili's ax was found in Arizona in 1966, a skeleton of a carnivore was also found that had two spearheads embedded in the skull, similar to Sili's axe, the stone had been sharpened with very precise cuts. Then in 1967 it was said that human bones had been found in a vein of silver from a Colorado mine. A 10 cm long copper arrowhead accompanied them. There was general agreement that the silver deposit was millions of years old and naturally much older than humanity.

Ancient hunting artifacts. In 1894 and 1895, scientific journals announced the discovery of cut-stones, that is, "worked" stones, from the Miocene in Burma, part of British India. The implements were reported by Fritz Noetling, a paleontologist who headed the Geological Survey of India in the Yenangyaung region of Burma. These fossils and tools were dated between 5 and 12 million years ago. Another interesting case occurred in 1870, when Anatole Roujou reported that the geologist Charles Tardy unearthed a flint knife from the exposed surface of the Late Miocene conglomerate at Aurillac, southern France. The French geologist JB Rames doubted that the objects found by Tardy were evidently man-made. Ergo, in 1877 Rames made his own finds of flint tools in the same region, at Puy Courny, a site near Aurillac. Those implements were taken from sediments laid between layers of volcanic materials established in the Late Miocene, which are 7-9 million years old.

Consider also the case of 1910, when Henri Breuil found stone tools in Belle-Assise, near Clermont, France. The formation in which they were found belongs to the Early Eocene (50-55 million years). Human tools were also found in Portugal. Carlos Ribeiro discovered flint implements in Miocene formations near Lisbon, Portugal. His discoveries were cited in several important books and significant archaeological records. In 1857, Ribeiro was appointed in charge of the Geological Survey of Portugal (Geological Survey of Portugal) and elected to the Portuguese Academy of Sciences. Most of their finds representing human evidence from prehistoric times are between 5 and 25 million years old.

Axes. We know that some thirty accumulated corpses were found in a place 50 m deep -in a place where no activity takes place-that they have chosen on purpose to accumulate these corpses. Along with those thirty bodies, they have deposited a biface, very well carved, of a strange color, rare in the materials found in the area. Baptized as Excalibur, this ax recently discovered in the Sima de los Huesos, in Spain, is around 400,000 years old. We also have the case of 1982, where KN Prasad of the Geological Survey of India reported the discovery of a crude pebble tool, a single-edged ax in the Miocene Nagri Formation near Haritalyangar, in the foothills of the Himalayan valley of the northwestern India. This places the ax in a remote antiquity of between 9 and 10 million years, being that, according to the accepted theory, the first tools would only have begun to be designed by the Homo habilis of Africa less than 2 million years ago.

Did Thor's hammer exist? The so-called "Texas Hammer" from the Somerwell Museum, found embedded in a cave in London (Texas) in 1934. It has been dated to 140 million years ago or more, and the iron is of great purity, only possible to obtain with modern technology, and the handle has accused a process of petrification in the rock, which shows its remote antiquity. This and the following evidence certify once again that there were men in the Triassic (Between 225 and 180 million years).

Mortar over 5 million years old. The Table Mountain mortar in California from 50 million years ago, and the Pestle and Pestle, which were discovered in 1877 when mining engineer JH Neale found several spearheads protruding from very dark rock in Toulomme, California. Examining these points more closely, he found not far away a small irregular mortar less than 10cm in diameter. At his side he discovered a pestle along with another more symmetrical mortar. Neal claims that it is completely impossible that these relics could have reached the position in which they were found at a time other than the formation of the gravel deposits, and before the lava layer formed. In addition, no traces of disturbance were observed in the rock mass or natural fissures that would have allowed the introduction of these objects through them. The position of these utensils found on Table Mountain indicates that they are between 33 and 35 million years old!

In 1853, a physician named Dr. HH Boyce discovered human bones on Clay Hill in El Dorado Country, California; In February 1866, Mr. Mattison, the principal owner of the mine at Bald Hill, near Angels Creek in Calaveras Country, dug a skull out of a layer of gravel 130 feet below the surface. The area where it was found coincided with many volcanic sediments and is much older than 5 million years, the same as in the case of Boyce. Stone tools over 5 million years old were also found nearby at the Smilow Mine and Marshall Mine in San Andreas, Calaveras Country, as well as at Spanish Creek in El Dorado Country, and at Cherokee in Butte Country.

In particular JD Whitney reported several discoveries coming from Placer Country. He mainly mentioned the discovery of human bones discovered in the Missouri tunnel: «*In this tunnel, under the lava, two bones were found [...] which Dr. Fagan pronounced to be human. One was said to be a leg bone; about the character of the other nothing was remembered. Most of the information was obtained from Mr. Goodyear from Mr. Samuel Bowman whose intelligence and truth-confidence the writer received good references from a personal friend well informed by him. Dr. Fagan was at that time one of the best doctors in the region.*» According to information provided by the California Division of Mines and Geology, the deposit from which the bones were taken is 8.7 million years old. Many such tools, as well as modern human jawbones, Neolithic tools, fragments of human skulls, carved stones, and other things out of time, ranging from 9-55 million years old, have been discovered at various locations in Tuolumne, Table Mountain in California on quite a few occasions.

Film cubes. As if it were the main plot of Steven Spielberg's film "Transformers", in the fall of 1885, a cubical object was found in a tertiary (60 million years old) block of coal in a mine in Germany. The parallelepiped was examined by Dr A. Gurlt. According to the 1886 publications on this object, it was first interpreted as a fossil meteorite and appears to be "worked, fabricated"!... The object measures 7 cm x 7 cm over 4.5 cm and its density is 7.75. We see that 4 of its faces are perfectly flat, and the 2 opposite ones are slightly convex. A deep groove surrounds him at half height. Something similar was also seen in England, when Osmond Fisher, a member of the Geological Society, discovered an interesting feature in the Dorsetshire terrain – the elephant ditch at Dewlish.

Fisher said in The Geological Magazine (1912): '*This trench was dug in lime and was 12 feet deep, and so thick that a man could simply walk through it. This does not fit the line of natural fractures, and the chert layers on each side fit [...] this trench, in my opinion, was dug by men in the Late Pleistocene era, and crafted with the intention of capturing elephants.*" Fisher made other interesting discoveries that he wrote about in 1912: "*When I was digging for fossils in the Eocene of Barton Cliff I found a piece of a jet-like substance about 9 1/2 inches square and 2 1/4 inches thick [...] the specimen is now in the Sedgwick Museum, Cambridge.*» Jet is a variety of lignite, a compact black coal that takes a good polish and is often used as jewelry. This cube was found in layers from the Eocene period dating to between 38 and 55 million years from the present.

Archaic currency. In 1871, William E. Dubois of the Smithsonian Institution reported many man-made objects found at deep levels in Illinois. The first consisted of a type of coin from Lawn Ridge, in Marshall Country, Illinois. The report said that in August 1870, at Lawn Ridge, near Peoria, Illinois, along with two companions, JW Moffit found a piece in the rubble of an artesian well they were drilling. Professor A. Winchell studied the object made of an unknown copper alloy at the time. Despite the corrosion, the round piece had very sharp and uniform edges throughout its thickness. A drawing that represented a crowned female face and looked like it had been etched with acid. On the other side, an animal with long pointed ears and a long frayed tail, was accompanied by another, similar to a horse. In the contour of the two faces some letters of unknown script could be seen. Found more than 30 m deep, it could be between 100,000 and 150,000 years old.

Moffit also reported that other artifacts were found in nearby Whiteside Country, Illinois. At a depth of 120 feet, workers discovered " *a large ring of ferrous copper, similar to those used on the yard (cross-sticks of the masts) of ships today [...] they also found something stylized-modified like a boot." -hook.*» Mr Moffit added: *'There are numerous instances of relics being found at shallower depths. A spear-shaped ax made of iron was found stuck in clay at 40 feet: and stone and ceramic pipes have been unearthed at depths ranging from 10 to 50 feet in various locations."* In September 1984, the Illinois State Geological Survey wrote to Michael A. Cremo and Richard L. Thompson saying that deposits to 120 feet in the Whiteside Country varied greatly. In some places, one would find 50,000-year-old deposits 120 feet away, while in others they find Silurian stones 410 million years old.

The pipes of Saint Jean de Livet. In 1968, Y. Druet and H. Salfati publicly claimed to have discovered semi-ovoid metal tubes embedded in Cretaceous limestone deposits 65 million years old, in a quarry in Saint Jean de Livet (France). After considering and rejecting various hypotheses, Druet and Salfati came to the conclusion that intelligent beings lived at the time attributed to the limestone in question.

Paleolithic bullets. One of the strangest things is discovering animals or men with bullet holes, such is the case of the skull of a Bison found in the Colorado desert, which had a perfect 2.54

cm hole, as if it had been killed with a laser beam of considerable power. Only the skull was from a prehistoric bison. One of the most spectacular finds in this regard is a skull that is currently in the Natural History Museum in London. It belongs to a Neanderthal man and was found near Broken Hill (Zambia) in 1921. On the left side of the skull is a round hole with flat edges. The cleanliness of the wound suggests that it was caused by a high-velocity projectile, such as a bullet. On the side opposite to this wound, the skull is destroyed as if by the action of the projectile leaving the skull. A Berlin forensic expert said the hole was identical to the gunshot wounds so often found today by men in his profession.

However, the remains were found at a depth of 18m. It was impossible for natural geological processes to cover it to such a depth even if the victim had died only a few centuries ago, since firearms first arrived in Central Africa after the 15th century. This skull that belongs to a Neanderthal curiously has a bullet hole made 40,000 years ago. It was analyzed by archaeologists and then by ballistics experts, unable to get out of their astonishment. Another case is that of a prehistoric animal shot, where an enigmatic object is added to many other similar findings. One of them being that of a skull of an aurochs (type of extinct bison) that was found near the Liena river - in the USSR - which presents a perfectly round and polished hole, similar to a bullet wound. The aurochs lived for many years after being injured.

The Coso Artifact. Baptized as "the Coso Artifact", in 1961 it was found in Olancha (California), near the Coso mountains, something surprising. The outer shell of the artifact is made up of hardened clay, stones, fragments of fossil shells, and two objects, resembling a nail and a disk. Inside, a ceramic cylinder can be seen in a hexagonal sleeve made of petrified wood, with copper fragments between the two. In the middle of the cylinder a 2 mm diameter metal stem is inserted. This object is between 250,000 and 500,000 years old.

The Dorchester Jar. A report titled: "A Relic of the Bygon Era" appeared in the Scientific American magazine of June 5, 1852, where it said that on Meeting House Hill, Dorchester (Massachusetts) an enormous mass of rock was unearthed, where some pieces weighed even tons. Among them, a metallic container was recovered in two parts, surely broken by the mining explosion. When the two fragments were joined together they formed a bell-shaped jar measuring 4 1/2 inches high, 6 1/2 inches at the base, 2 1/2 inches at the top, and about an eighth of an inch in the middle. The color of said vase was similar to zinc, although it seemed to have certain portions of silver. Around it were six floral figures that stood out for being made of pure silver and around a portrayed vine, also made of pure silver. This impressive artifact was pulled from a depth of 15 feet and remained in the possession of Mr. John Kettell. According to a recent study of the US Geological Survey map of the Dorchester-Boston area, the stone, now called the Roxbury conglomerate. It belongs to the Precambrian era, over 600 million years old.

Spoons, bells, rings and hooks. In 1976 a newspaper published the description of a spoon that was found in 1937 mixed with soft coal from Pennsylvania. The spoon was found in a mass of brown ash resulting from the combustion of a large piece of coal. When removing the ashes, the spoon appeared, which could possibly be a relic of the antediluvian world. In 1851, in Whiteside County (Illinois), during excavations two copper objects were taken from a depth of 36 m. They resembled a hook and ring about 150,000 years old. In 1944 Newton Anderson found a bell inside a lump of coal rock near his home in West Virginia. This artifact has been studied by the University of Oklahoma and shows to be hundreds of thousands of years old. The artifact has been coined the name "Adam's Bell".

Gold chains and threads. It is difficult to think that cavemen made gold threads. It's strange, but it happens, and 'The London Times', from June 22, 1844, reported that in Scotland - between the Tweed and Rutherford rivers - at a depth of 2.5 m, some workers found a gold thread embedded in a rock, which according to Dr. AW Medd of the British Geological Survey in 1985 belonged to the Carboniferous (320-360 million years). This thread was exposed at the headquarters of the local newspaper, the "Kelso Chronicle". On another occasion, in 1891, in Morrisonville (Illinois, USA), when breaking a large block of coal, Mrs. SW Culp found a gold chain about 25 cm long, the ends of which appeared caught in two different pieces. . This was reported by The Morrisonville Times on June 11 of that year. The Illinois State Geological Survey said the layer in which the gold chain was found was 280-320 million years old.

The millennial Nails. In the 16th century, in 1572, an 18cm iron nail was found in the rock of a mine in Peru. It was given as a souvenir to the Spanish Viceroy of Peru. The age of the geological layer from which it was taken is estimated to be between 75,000 and 100,000 years. In another case, in 1844, another inexplicable iron artifact was subjected to careful and detailed investigation. Sir David Brewster reported that a 60cm long block of stone from the Kingoodie quarry near Dundee, Scotland, which was being cleared, produced a rusty iron nail, found at the point where the stone and earth they were. The pointed end of the nail projected a little more than 1 cm towards the Earth, while the rest rested on the surface of the stone, except for the last 2.5 cm of the head end, which was driven into it.

The block was estimated to have formed 60 million years ago, although Dr. AW Medd of the British Geological Survey wrote in

1985 that the sandstone belonged to the "*Old Era Lower Red Sandstone*", i.e. Devonian (360 and 408 million years). Brewster was a famous doctor who made important discoveries for the British Association for the Advancement of Science. Another similar case happened a few years later, when in 1851, in Springfield (Massachusetts), Mr. De Witt accidentally broke a piece of gold-bearing quartz that he had brought from California. Inside was a 5cm wrought iron nail, straight with a perfectly formed head and slightly corroded. The stone was 1 million years old.

Old screw. Called the Lanzhou screw, it was discovered in China's Mazong Mountains in June 2002. The rock containing the body of the screw was found by Mr. Zhilin Wang during a field investigation in the intermediate zone between the provinces of Gansu and xijiang. The color of the rock is an unusual black, and its degree of hardness also makes it particular. Its weight is 466 grams and its approximate dimensions are 7 x 8 centimeters.

The object inserted in the rock presents all the characteristics of the body of an ordinary screw, about 6 centimeters in length. Since its appearance, the screw body has attracted the attention of many scientists and researchers, from institutions such as the "Gansu Province National Bureau of Land Resources", the "Research Institute of Geology and Minerals", and the "Lanzhou University School of Resources and Environment". After several investigations, the specialists confirmed that the rock should be treated as one of the most valuable objects of Chinese and world archaeology. On another occasion we also found that, in 1865, a piece of feldspar, extracted from a mine in Treasure City (Nevada), contained the rusty remains of a sharpened screw. The stone was 21 million years old. What dinosaur made screws?

Eve's Thimble. Around 1880, in the state of Colorado, a rancher went out to look for coal from a seam that existed on the side of a hill. The cargo it picked up came from a location about 45m from the mouth of the vein, and about 90m below the surface. Returning home, he began to split the pieces of coal, and from one of them an iron thimble jumped out. Or at least, it resembled a thimble, and was soon known locally as "Eva's Thimble." It had the same notches that modern thimbles have. The metal crumbled into crumbs as it was handled by curious neighbors, until it was finally lost. Even admitting that the Indians used iron thimbles in remote centuries, the mystery remains, since the coal from which this object came was formed between the Cretaceous period and the Tertiary era, some 70 million years ago! And according to the opinion of the experts, humanity did not yet exist, but rather the closest thing to human beings were small lemur-like mammals that lived in trees.

Laon's Ball. In April 1862, The Geologist published a report documenting the discovery of a limestone ball at a depth of 75 m in lignite beds near Laon and belonging to the Tertiary Period. Various fossil forms were visible in the layer of sandstone clay that was on top of the lignite. It was in August 1861 when miners working at one end of the tunnel saw a round object fall from the top of the excavation. The spherical object was about 6 cm in diameter and weighed about 310 grams. According to Maximilien Melleville, vice president of the Laon Academic Society and author of the report, there is no doubt about the authenticity of the sphere. In their assumptions, logically, there was no place for the possibility that man had existed when the lignites of the Paris basin were formed. If this ball was human work, Melleville's reluctance would be justified, since it would mean admitting that some 50 million years ago (Tertiary Era) an intelligent culture inhabited France.

The Carboniferous Iron Cup. Other evidence from the Carboniferous (Between 345 and 280 million years ago) is the iron cup in Oklahoma, USA. On January 10, 1949, Robert Nordling sent a photograph of an iron cup found in 1912 inside a piece of coal belonging to the mines of Wilburton, Oklahoma, to Frank Marsh, from Andrews University, in Michigan (USA). The photograph belonged to a friend of Nordling's, whose father worked at the Thomas Municipal Power Station, Oklahoma. The discovery occurred when a company employee worked next to the boiler. Upon hitting the coal, the fragment inside which the cup was found broke, exposing it. According to Robert O. Fay of the Oklahoma Geological Survey, the coal from the Wilburton mine is about 312 million years old.

When Frank Marsh received the photograph he tried to find the owner of the cup, but his efforts were unsuccessful because Nordling's friend had died and his heirs did not know the whereabouts of the mysterious utensil. Years later, in 1966, Marsh began a series of contacts with Dr. WH Rusch, a biology professor at the Concordia School in Ann Arbor, Michigan, in order to clarify the mystery, without success in his inquiries. Be that as it may, it is a real tragedy not to be able to count on this precious relic, since it would undoubtedly help to clarify some of the enigmas that the most daring archaeologists face.

Stones with metallic inlays. From elrisco.wordpress.com, there is an article claiming that archaeologists from the University of Saint Petersburg have certified the authenticity of some fossils found 200 kilometers from Tigil, on the remote Kamchatka peninsula, Russia. The paleontologist Yuri Golubev tells surprised how the finding can change history as we know it so far. It is not the first time that OOPARTS (Out of Place Artifacts or Artifacts out of place) have been found, but it is surprisingly preserved. It is a rock that contains an infinity of metallic pieces. These pieces seem to form some mechanism of gears that perhaps belonged to a clock or computer. The surprising thing is that they are fossilized in a rock that is no more and no less than 400 million years old.

They say they received a call from the mayor of Tigil who said that some hikers found the remains in a very steep area. When the investigators moved to the place, they did not believe what they saw. There were hundreds of little wheels that seemed to make up a machine. They were perfectly preserved, as if they had petrified in a short period of time. They said that they had to put surveillance in the area because the visits of curious people have multiplied by a hundred: <<*The other day several groups of North American geologists arrived and although we cannot deny them to pass by, we do keep a close eye on them.*>> Nobody could believe that 400 million years ago there was a man on Earth. In fact, the ways of life back then were very simple. This points to a probability that the beings that brought that technology came from outside, or that there were more advanced civilizations in our past that died out. The Ancient Astronauts theory would hold that they may have been visitors and their ship was damaged and they had to stay. With the passage of time, they would very possibly begin to develop technology with their knowledge and thanks to the materials found. Another possibility is that maybe they were parts of a computer that they used to calculate a route, like the famous "Antikythera Mechanism."

From the tests carried out so far, it has been found that the pieces became fossilized in a short period of time. These fell into a quagmire that became fossilized due to a strong cataclysm. Denying the existence of technology in the past is a serious mistake, because evolution has not been linear. It is known that there have been major cataclysms and many of these have wiped out intelligent life on the face of the Earth on various occasions. Yuri ends the short interview with the following sentence: *"Our current level of technology compared to this is proportional to the ability of some to invent things ..."*

The iron measuring cup. Among the great researchers of these OOParts (Artifacts Out of Time) with, for example, Klaus Dona, Richard Thomson, Michael Cremo or Dr. Zillmer who studied a kind of iron measuring cup discovered in 1912 when it fell to the sea split a large piece of coal. The coal came from the mine in Wilburton, Oklahoma and appears to be over 300 million years old. It is exhibited at the "Creation Evidence Museum" in Glen Rose, USA.

Kentucky and the age of the trilobites. In Kentucky, USA, there is today a museum where impressive pieces are preserved, whose age and material astonishes us:

- Fossils of some kind of reptiloid of millions of years are found.
- There is a petrified bone with fluorescence, which in itself is unusual.
- There is a sarcophagus with an image of a human face.
- Contour of petrified hands from the age of dinosaurs.
- A carved tetragonal metal which has the modest antiquity of 4,000 million years.
- Part of a giant petrified body with something strange on it, which resembles the emerald.
- A petroglyph of millions of years.
- A petrified trunk carved in the Carboniferous.
- An unknown metal alloy from Burnside, Kentucky, discovered 300 years ago.
- An unknown metal artifact that was kept at the Huntsville Space Center, and whose color is like a mixture of copper and bronze.
- There is a frozen fossilized artifact.
- A fossilized artifact found with human finger marks, which cannot be distinguished if it is stone, glass or metal, but its color seems to be an artificial yellow.
- Another is a possibly meteorite artifact, but with unknown engravings, dating back some 4 billion years.
- There is also a projectile head preserved in quartz of millions of years.

- There's a golden nail inside quartz, also millions of years old.
- The other thing like that, remarkable, is a stone in the shape of a huge heart, which is 300 million years old, and in its center there is a perfectly carved hole.

Klerksdorp spheres. Many science and archeology magazines have echoed in recent years the discovery made by several South African miners in the town of Ottosdal (South Africa). Some spheres were found among pyrophyllite, a type of material that is very soft and is formed by sediments of more than 2.8 million years. Roelf Marx, director of the Klerksdorp Museum (South Africa), where some of these spheres are kept, says: «*the objects appear artificial, but the rock stratum where they were found corresponds to an era in which no intelligent life form existed. I have never seen anything like it.*" The size of these spheres ranges from 3 cm to 8 cm in diameter, some of them housing a spongy material inside that vanishes very easily when sectioned and left in contact with air.

Its exterior is formed by a highly hard alloy of steel and nickel, powerfully drawing attention to some fine lines or grooves that surround the spheres, dividing them into two equal parts. These stones can be divided into two types, the first are a solid bluish metal with white spots and the second are hollow and filled with a white spongy material. On the other hand, A. Bissehoff, professor of geology at the University of Potchefstroom declared that: "*the spheres were made of an agglomerate of limonite, a type of ferric material.*" Bissehoff's hypothesis is not accepted by the international community since limonite is a relatively soft metal and would admit scratching by steel, which is not possible since the spheres are much more resistant, in addition to the fact that limonite agglomerates appear in groups and never isolated nor so perfectly spherical. It would be like admitting that they were carved with something similar to a laser, without leaving cracks or traces of splintering, typical of tools. The find speaks for itself.

Concrete walls of the Devonian age. We have in the list the case of a block wall in Oklahoma (USA), where WW McCormick of Abilene (Texas), reported that in 1928 Atlas Almon Mathis was working in a well 2 miles from Heavener, Oklahoma, when he ran into several concrete blocks that were scattered on the ground after a series of detonations that were caused to open the mine. The blocks, 30 cm on each side, were smooth and polished, so much so that they looked like a mirror. At a depth of 100 or 150 m, another miner found another similar wall. The age of the coal deposited in this mine is estimated at 286 million years. After this incident, the mine was closed and its workers were ordered to keep quiet about such an amazing find.

Some researchers, such as the historian Michael A. Cremo and the philosopher of science, Dr. Richard L. Thompson (writers of the famous book "Forbidden Archeology"), could not verify the existence of the find, but managed to unearth many other testimonies from miners. that during their days working underground they came across vestiges of the remote past that barely matched the official chronologies. One of these stories refers to a miner named James Parson, who, working with his two sons, came across a slate wall in a coal mine in Hammondville, Ohio, (USA), in 1868. According to an article Published by MK Jessup in 1973, the wall in question was smooth and engraved with several straight lines of raised hieroglyphics.

As a result of this event, according to Atlas Almon Mathis, a worker at this mine, they took us all out of there and forbade us to speak about it, later we were moved to mine number 24, near Wilburton, also in Oklahoma. In this mine some of Mathis's companions found a solid block of silver in the shape of a barrel with the stave mark (boards that form the sides of a barrel) on it. The Wilburton coal formed 280 to 320 million years ago.

Baiun Tunnel. Another equally fascinating construction is the "Bainun Tunnel" in Yemen. Attributed to Solomon and the Queen of Sheba, it is another of the great paradigms of the past, and one of the many buildings that the Israelite King Solomon ordered to be erected. "Gundam", a famous castle built just after the Flood, and which Yemeni archaeologists believe could be found very soon. This, together with the Castle of Salín, are attributed to the "geniuses" (fantastic beings) of Solomon. The Arab historian Al-Hamdani assured in one of his various works that he himself had seen the castle with his own eyes. This must have gone back to around 930-940 AD. C. and his statement coincides with his Afghan colleague Biruni, who saw the ruins of Gundam in the city of Sana, at the foot of a mountain called Nikkum or Lokkum, and says that this city of Sana was built by Sem, the eldest son of noah.

Strange carved stones. On April 2, 1897, the Daily News of Omaha, Nebraska, (USA), published an article entitled: *"Carved stone buried in a mine."* The article said that while miners were working in the Lehigh coal mine in Iowa, at a depth of 40 m, one of them found a piece of rock that did not correspond to the surrounding rock. The stone in question was dark gray and measured about 60 cm long, 30 cm wide and 10 cm thick. On the very hard surface of the stone there were several drawn lines that formed perfect rhombuses. In the center of each of these diamonds was represented the face of an old man. How did this stone get there? The miners who found it insist that the Earth in that area had never been worked in prospecting. An authentic enigma starring a coal belonging to the Carboniferous Age (Between 345 and 280 million years).

One of the most famous anachronistic objects is the one known as the "Salzburg Cube" found in 1885, when a worker at an Austrian iron foundry was breaking chunks of Wolfsegg coal, he found an iron object with a cubic shape, although somewhat deformed. Noorbergen repeats the description of the object, which soon became well known: The edges of this strange object were previously perfectly straight and defined; four of its sides were flat, while the remaining two sides, located opposite each other, were convex. Halfway up it had a fairly deep groove.

Actually, the shape of the object, which is now in a municipal museum near the foundry where it was found, does not resemble a cube at all: its only flat surface is the result of a slice that was separated from it for analysis. chemically. The analysis showed that the metal does not contain nickel, chromium or cobalt, so it cannot be a meteorite, as had been thought at first. It looks like some kind of wrought iron. The crucial question is whether it really formed within a lump of coal. It seems that the scientist who first investigated the cube and suggested that it was a meteorite did not even try to find the lump of coal with the cavity that had housed the cube. In the absence of this crucial piece of information, the Salzburg cube received publicity out of all proportion to its intrinsic value.

The Ica Stones. Much has been said about these stones, mixing the frauds and the stones carved by villagers to sell as artisan pieces, with the true volcanic stones with prehistoric carvings. The French Chanoux, in his work "Enigma of the Andes", assured that the stones of Ica could be "*the library of the Atlanteans that have existed 50 million years ago.*" One of these enigmatic stones portrays the world in the Paleozoic and Lower Mesozoic, based on a supercontinent called Gondwana, and whose capital was Panguea -the current Caribbean, and means: "All Lands"-, and whose mega ocean was "Panthalasa". (The "Sea of Tethys", central sea, was in the Gulf of Pangea. This gliptolithic humanity decided to fix their knowledge in stone (and other materials such as precious metals, destroyed by human greed) to avoid catastrophes for future men and help them govern their lives according to wise and rational rules. One of the first peoples to do so was, according to Dr. Cabrera, the Inca people.

Stones are observed where prehistoric men perform operations of all kinds on other men and women. One of the elements that confirm the belief of Dr. Cabrera, who died in 2001, is a stone where a map of the world is carved, just as it was in the tertiary period. There, the shape and layout of the continents is completely different from today, "some areas seem to coincide with the missing continents of Lemuria and Atlantis." Considering that geology did not know until the end of the 19th century and the beginning of the 20th that the great cataclysms of the end of the Tertiary had caused spectacular changes in the shape and arrangement of the continents, Dr. Cabrera maintains that this stone could only have been carved by men who They lived on a planet with this configuration and, furthermore, they had the necessary technical means to explore it and observe it from great heights.

Years ago these stones were analyzed again and one of the two studied gave a date of 60,000 years. The natives always buried themselves with them since at least 7,000 years ago; they were considered stones of the gods. These magnificent evidences of the past are spread over 1/4 of Peru. Many people reburied them without knowing what they were. They have appeared in rivers, valleys, slopes, etc. The Ica Stones, prehistoric prints that shed much light on this truth that frames our past, were first studied by Dr. Cabrera, who named them "glyptoliths" and describes those who carved them as "gliptolithic humanity". Based on his interpretations of the drawings engraved on the stones, Cabrera affirms that this "gliptolithic" humanity was created by a superior race that arrived on Earth from somewhere in the cosmos.

Upon reaching our planet, that race found no intelligent life, and decided to create it from a primate related to the lemur, called "notharcus", which became extinct 50 million years ago. In his book

"The message of the engraved stones of Ica" (Inti Sol editores, Lima, 1976), he states: "*By transplanting cognitive codes to some primates that belonged to a type of highly intelligent primate, they generated men.*" Apparently, the stones say that there were several human categories: those with the greatest cognitive power. These are what Dr. Cabrera calls "reflective and scientific men", above which, of course, their creators, men from the cosmos, ranked. It was said that Dr. Cabrera had forged a large number of stones, but it was following the models already found since colonization. It turns out that there are chronicles of priests from the colony that describe these stones, proving that they are not a fraud.

They exist, and the real ones have been added to the falsifications, but in the Spanish chronicles they are already talked about. The stones were sent to a respectable mineralogy laboratory for analysis and they determined that the incisions made to make the figures are covered with an oxide that could only have formed, not in weeks or months, but in thousands of years. In addition, the people who were selling them illegally were ignorant who made believe that they were the ones who carved them so as not to go to jail for selling a nation's heritage. Likewise, the people who were forging these stones, to manually carve a single and small size, took approximately 5 months. How were they going to carve 11,000? Apart from this, there are even larger carved stones. Those did not have the same preciousness or perfection as the originals, and the sellers used any type of stone.

The originals are andesite-type stones, which are one of the hardest known and are not found on the coast, but in the bowels of the Andes. Who carried that to that area? Let's put this idea in mind: these stones show operations that cannot be performed even today and comet steps that were not known until 40 years ago. The book "Las piedras de Ica" by the Spanish writer Juan José Benítez, explains that in Peru there are about 11,000 stones out of 40,000 that are believed to exist throughout the Earth, and that they are shown or "recounted" with laser carving about them, the story of a humanity that not only lived on this planet but also took part in the extinction of the dinosaurs, as well as many other things in prehistory. This explains why a considerable number of these dinosaur fossils have been found with holes from what appear to be "projectile" or "laser" holes in the back of the skull.

It is common to see a sauropod attacking a man or several carnosaurs tearing a man to pieces on the stones. Dr. Cabrera, based on the fact that the engraved stones are geologically andesites, that is, stones that were formed in the tertiary period, affirms that it was in that [prehistoric] period when the superior beings that arrived from space created humanity. So far, analyzes have not confirmed that the engravings are strictly contemporary with the stones; however, some microorganisms found in the grooves of the engravings do date back millions of years. On the other hand, there are other indications, in Latin America itself, that point to a greater antiquity of man.

The spheres of Costa Rica. Like the hundreds of spheres that have been discovered in Gongxi Town, in the Hunan Province of China, and which are in the Shennongjia Nature Reserve Area, in Hubei Province, there are also exactly a certain number of spheres in Costa Rica. ancient stone. It was during the 40's, when a North American banana company began its exploitation in the Diquis delta, in the southwest of Costa Rica, when, when beginning the work of clearing the forest, preparing it for cultivation, some imposing rocky stones of different sizes and spherical shape. They are of variable size: the smallest are only a few centimeters in diameter and the largest spheres have a diameter of more than 2m, the latter weighing up to 16 tons. They are built in granite stones, andesite and sedimentary rock.

It is believed that the stones were transported by the river from many kilometers away to their current location, since these types of stone have not been found in the Diquís delta area. Although most of the spheres are found in pre-Columbian archaeological sites, there is no way of knowing if they were made by them or by some other culture prior to this one. Immediately after its discovery, the archaeologist Doris Stone carried out a series of investigations that were in vain since she was unable to date the antiquity of the stones, with what tools they were so perfectly made, nor their origin. Subsequently, Samuel K. Lothrop, an expert in indigenous civilizations and archaeologist, set out to unravel the enigma of these spherical stones, but could not formulate any conclusive theory. More recently, groups of archaeologists have investigated the Diquis spheres with more modern methods, reaching the conclusion that they began to be made about 3,000 years ago.

There is the theory of astronomical representation. In this hypothesis (disclosed by the researcher Michael O'Reilly) the stones are identified as possible celestial charts with a ceremonial purpose or as an indicative calendar. In 1979, one of these stones was found in Guayabo de Turrialba (Cartago province), which could have had the function of a precision calendar, and which, together with the use of small astronomical objects, gave details of dates such as the solstices. , the longest day of the year and the duration of the rainy season. Another unconventional theory, carried out by the Estonian anthropologist Ivar Zapp in his book "Atlantis in America", affirms that the stones could originate from Atlantis, an island-continent that disappeared 12,000 years ago, and although the archaeological authorities of Costa Rica did not strongly agree with this theory, the International Biographical Center mentioned Zapp as one of the most renowned scientists of the 20th century.

Iván Zapp discovered with the help of Carlos Araya (Commander of the Costa Rican Airlines) and an atlas, normal at the beginning and Mercator (atlas that takes into account the curvature of the Earth) later, that the spheres exactly where they were located when they were discovered, they pointed in different directions, just as if they were large-scale maps. One of the alignments unearthed by archaeologists showed the path in a straight line that leads to Cocos Island, then to the Galapagos Islands and finally to Easter Island. A second group of rocks pointed to the islands of Jamaica, Cuba and Bermuda. While others were oriented towards Giza, in Egypt and Stonenhenge in England. Thus confirming that they were routes to other parts of the planet.

CONCLUSION

THE LAST HUMAN TRANSITION

Although we cannot, at this point, say that man appeared on Earth recently. Some will wonder if perhaps we came from Adam and Eve as the Jewish, Christian and Muslim religions say. The truth is that Adam and Eve were probably Caucasians who appeared on the scene about 9,000 years ago, but we have seen that there is incredibly superior human evidence. We will deal with the story of Adam and Eve in future editions due to the large amount of material on the subject. For now, we will finish this part to give rise to the unknown history of our planet in our works of "The Sakla Rebellion". Neither chance nor improbable mutations gave rise to the human being, something else did.

Authors Max H. Flint and Otto O. Binder wrote in their book, Mankind, Child of the Stars: *"Cro-Magnon man appeared with mysteriously enhanced skeletal features, and with an ability which is*

315

astonishingly in excess per 100 cubic centimeters of that of modern man... A similarly large degree of brain expansion occurred in absolutely no other species on Earth in all past ages, nor has any gene shown evidence of brain mutation of comparable magnitude since ancient times."

Alan Alford, one of many scientists who are open to the concept of extraterrestrial intervention, explains the anomalies surrounding the origin and supposed evolution of man: «*Homo sapiens have acquired a modern anatomy, language capacity and a sophisticated brain (much more beyond the necessities of their daily existence) apparently in defiance of the laws of Darwinism. There are a number of possible explanations for this anomaly. One is that humanity evolved in the sea, and that crucial fossil evidence is missing. Another is that the Darwinian Theory itself has a Missing Link. And a third explanation is that the genes of modern man were suddenly implanted by an intelligent extraterrestrial species that colonized the Earth."* In his book, "The Neanderthal Enigma", James Shreeve comments on the enigma of our human development: «*A 'major transition' did occur, but it happened so close to the present moment that we are all still reeling from it. Some here in the hall of history, just before we started keeping records about ourselves, something happened that turned the passably precocious animal into a human being."*

Maverick catastrophist scientist/cosmologist Immanuel Velikovsky, probably one of the most abused scholars in history, wrote: "*The most controversial is the evolutionary question. I have done a lot of work on Darwin, and I can say with some certainty that Darwin did not derive his theory from nature, but rather superimposed a certain philosophical view of the world on nature, and then spent 20 years trying to collect facts for make her fit.*" Researcher Laurence Gardner puts the impossible process this way: "*It took man over a million years to progress from using stones as he found them, to the realization that they could be carved and shaped for a better purpose.*

Then it took another 500,000 years before Neanderthals mastered the concept of stone tools, and another 50,000 years before grains were cultivated and metallurgy was discovered. So, by all the scales of evolutionary estimation, we should still be far from any basic understanding of math, engineering, or science, but we're only 7,000 years later, sending probes to Mars... So how do we inherit the wisdom and whose?»

Max H. Flint and Otto O. Binder, these acclaimed scientists, point out that there are many other specialists with serious doubts about the Evolutionary Theory Relationship of the origin and ascent of Man. They state: "... *we are not arguing with natural selection...since it applies to other creatures...But we do state unequivocally that Darwinian Evolution and natural selection do not apply to Humanity at all.*" They call our attention to the date 35,000 years BC, to the supposed species-specific transition from the Neanderthals -who we have already seen were not our ancestors-, to the Cro-Magnons, who are believed to be, although their bony characteristics, themselves, differ considerably from ours. They write: "*And the biggest riddle of all, where did Cro-Magnon Man, the first of our own species, Homo sapiens, come from 35,000 years ago? The disappearance of Neanderthal Man and the arrival of Cro-Magnon Man at approximately the same time is one of the truly great stumbling blocks for the theory of Evolution, since these are non-sequitur species. The Neanderthal had a large brain, but a small mental capacity. He was a mass of muscle. This man was suddenly replaced by the Cro-Magnon, who was an entirely separate species.*"

They argue that "... *the Neanderthal most decidedly could not be the direct ancestor of the Cro-Magnon, since they were two different types of humans, physically and even skeletally. The Neanderthal man endured cold cycles and warm cycles successfully, it seems. He continued to exist in Western Europe until about 35,000 years ago, and then abruptly disappeared. The evolutionary trends that it exhibited during this period are extremely puzzling, as it seems that it became more 'primitive' rather than less. The classical theory of Evolution simply cannot explain these two events. First, the precipitous demise of a well-established species, plus the abrupt debut of a new species. Second, the fact that the Neanderthal species regressed and became more primitive over time. Natural selection and survival of the fittest are square nails that cannot be inserted into those round holes ..."*

These two scientists are convinced that biogenetic intervention and hybridization led to the extinction of the Neanderthals and the sudden appearance of some much more sophisticated Cro-Magnon. And they are far from being speculators, in the orthodox academic community. Dr. Robert Bloom, a renowned paleontologist, came out with a statement that probably stunned all his colleagues, saying that it was clear to him that what counts as evolution was achieved not by natural selection or mutations, but by the intervention of "... *spiritual beings in various degrees and of various kinds of intelligence.*"

To be continue...
... with the saga 'The Sakla Rebellion'.

EPILOGUE

As I mentioned at the beginning of this work, Creation cannot, no matter how hard it tries to justify it, be the product of chance. To visually understand how the universe is truly a hologram, a Yale University mathematics professor developed a formula and entered it into a computer program. Named after his name "The Mandelbrot Set", it shows an apparently disordered pattern, but by increasing it regardless of how close you get to the design, we will always find the same pattern within itself. Each "fractal" broken down infinitely always reflects the totality of it. When a fractal changes its pattern, the sum total of the entire pattern changes throughout the pattern.

« *This means that the whole world does not need to be awakened, there is no need to inform the 6 billion people of the planet of this message. It is only important that, on a personal level, you learn to conquer the fears within you and learn to love, when you see the order of your fears and ignore your emotions, then and only then will you be truly free.* " (Esoteric Agenda) In our next issue we will discuss more information about human prehistory, unknown races, lost technology and the events that led man to the caves having had a great heyday and knowledge, even more than what is estimated. We will also deal with the mythical rebellion of Lucifer, the evidence of other races in our universe and the birth of the Golden Age of the great gods: «... *those who believe in such visits to Earth are contemplating the possibility of other galaxies or other stars distant as* *the home of these <u>extraterrestrial astronauts</u>.*" (Zecharia Sitchin. The Twelfth Planet, 1976)

Main Sources

B**ooks and documents:**
Harun Yahya: "The Deception of Evolutionism" - book and documentary.

Richard L. Thompson and Michael A. Cremo: "Hidden History of the Human Race" (1999) and "Forbidden Archeology".

Zecharia Sitchin: "The Twelfth Planet" (1976).

RA Boulay: "Dragons and Flying Serpents" (1990).

Sixto Paz Wells: "The Cosmic Plan".

William Bramley: "Gods of Eden" (1996)

Michael Tsarion: "Atlantis, Alien Visitation & Genetic Manipulation" (August 2002).

Louis Pauwels and Jaques Bergier: "The Return of the Sorcerers".

Erich von Däniken: "We are all Children of the Gods".

David John and Richard Moody: "The Prehistoric World" (The World of Knowledge, Ediciones Vidorama, SA - Colombia).

National Georaphic: television network and magazine.

Very Interesting (magazine).

The Hebrew "Torah", Greek versions of the Septuagint and other passages from the Reina y Valera 1960 Bible.

Documentaries and conferences:
Michael A. Cremo: television interviews.

Harun Yahya: TV interviews.

Odyssey: "The Greatest Hoax of the 20th Century".

BBC: "The Day The Earth Nearly Died".

Carl Sagan: "Cosmos."

Klaus Dona: "The Hidden History of the Human Race" – interview.

Gregg Braden: "The Science of Miracles."

Drunvalo Melkizedek: "Secret Planetary History".

Félix Guttmann: studies and conferences.

Michael Tsarion: "Architects of Control" and interview on Sci-Fi.

Documanía: "The Corporation".

Iberian America: "Mysteries of the Past: The Biblical Flood" (Akásico 2004).

"So, what do you know? Inside the Burrow."

"Esoteric Agenda" (Talismanicidols.com).

History Channel: "Ancient Anstronauts."

Extranormal: "Intraterrestres", Mexican television network.

Networks, 2 Spanish TV: "Charles Darwin and Evolution."

Salfate: "Así Somos" Chilean television program.

Websites:

projectmagen.com

caminoluz.org

Allforjesus.creatuforo.com

The-inexplicable.com.ar

Messianic Wars (blogspot)

akasico.com

Forocristiano.iglesia.net

harunyahya.com

Wikipedia.com

Alberto Canosa (blogspot)

projectcamelot.com

blackvault.com

Disclosureproject.com

Darwin's Laboratory (blogspot)

Others:

Al Gore: "An inconvenient truth" (Ed. Gedisa. March 2007)

Alexander King and Bertrand Schneider: "The First Global Revolution" (Club of Rome. Pantheon Books. New York. 1991)

Allianz AG: "Demographic Change: Population Growth" (July 21, 2007)

Spanish Association of the Electrical Industry: "Environment Bulletin". (October 2007. Number 86)

Valencian Association for the Defense of Life: "Provida Press" (Number 224. Valencia. June 26, 2006)

United States Agency for Development (USAID): "US Private and Voluntary Organizations Registered with USAID"

BK Skinner: "The Gumption Memo" (Fall 1996)

Cardinal Newman Society for the Preservation of Catholic Higher Education: "The Culture of Death on Catholic Campuses" (Manassas, Virginia. April 2004)

Center for the Study of Carbon Dioxide and Global Change: "Testimony of Al Gore before the United States Senate Environment & Public Works Committee" (Washington, May 2007)

Bioethics Center of the Catholic University of Córdoba: "Gacetilla 09/08" (Córdoba. Argentina. June 26, 2008)

Newspaper "Get Real": "Al Gore in Lisbon" (Lisbon. February 13, 2007)

Newspaper "Latino" (Madrid. Friday June 22, 2007)

Focal Point for Women in the Secretariat: "NETWORK—The UN Women's Newsletter" (New York. Volume 11. Number 2. April, May and June 2007)

United Nations Population Fund: "Actions to achieve sustainable and equitable development" (New York, 2002)

Haddam-Killingworth High School: "The Cougar Chronicle." (Higganum, Connecticut. December 2007. Vol. 16 Issue 3)

Italian American Journal (New York. June 2008)

John Clapper: "The sanctity of human life and abortion" (WRS Journal 5/2. Tacoma, Washington. August 1998)

Joseph H. Hulse: "Sustainable Development at Risk: Ignoring the Past" (International Development Research Centre. Ottawa, Canada. Cambridge University Press India Pvt. Ltd. Year 2007. ISBN (e-book) 978-1-55250-368- 3. ISBN 978-81-7596-521-8)

J. Trigo: "Sustainable Development: Does it serve to reduce Poverty?" (Participation in Conference organized by Fundación Iberdrola

Madrid, December 18, 2002)

Miguel Zafra: "Al Gore's 9 errors" (Living History. Bulletin of the Social Area of the Virgen de Europa College. Madrid. November 2007)

Online US News: "Approval of abortion drug changes medicine and politics of issue" (September 10, 2000)

Optimum Population Trust (OPT): "The Jackdaw" (London, August 2008)

Optimum Population Trust (OPT): "Too many people: Europe's population problem" (London. August 31, 2008)

International Socialist Organization: "International Socialism" (San Juan, Puerto Rico. Year 6, No. 33. June 2003)

Paul Detrick: "Definitely Some Inaccuracies in Gore Film" (News Busters. October 4, 2007)

Population and Sustainability Network: "Population Matters" (London. Newsletter No.1. April 2004)

Rotarian Action Group for Population & Development: "Fragile Earth" (Lawrenceville, Georgia, USA. March 2007)

Servando González: "Evolution in the revolution? A response to Carlos Wotzkow" (Xzault Media Group. California. July 2007)

The Conservative Party of New York State: "National Affairs Platform" (Brooklyn, NY. September 6, 2008)

US Information Agency: "Population at the Millennium: The US perspective " (Global Issues Magazine. Volume 3. Number 2. September 1998)

Wanda Franz: "Social Justice and Wantedness" (Speech at the Catholic Press Association Convention. Baltimore. May 26, 2000)

God bless you!
Frederick Guttmann R.
frederickguttmann@gmail.com
www.frederickguttmann.com[1]

---

1. http://www.frederickguttmann.com

Sovereign State '*Eretz Yeshua, REML*'
Non-Governmental Organization "*Energy Angels*"
Non-Governmental Organization "*Misión Mesiánica Mundial*"

Misión Mesiánica Mundial

2002 – 2012 ©

About the Author

Israeli writer, researcher, disseminator, documentary filmmaker and influencer. He is the writer of more than 35 books, mostly research and dissemination theses.

Read more at https://www.frederickguttmann.com.

www.ingramcontent.com/pod-product-compliance
Lightning Source LLC
Chambersburg PA
CBHW021939120726
47992CB00001B/49